Selected Titles in This Series

12 **Salma Kuhlmann,** Ordered exponential fields, 2000
11 **Tibor Krisztin, Hans-Otto Walther, and Jianhong Wu,** Shape, smoothness and invariant stratification of an attracting set for delayed monotone positive feedback, 1999
10 **Jiří Patera, Editor,** Quasicrystals and discrete geometry, 1998
9 **Paul Selick,** Introduction to homotopy theory, 1997
8 **Terry A. Loring,** Lifting solutions to perturbing problems in C^*-algebras, 1997
7 **S. O. Kochman,** Bordism, stable homotopy and Adams spectral sequences, 1996
6 **Kenneth R. Davidson,** C*-Algebras by example, 1996
5 **A. Weiss,** Multiplicative Galois module structure, 1996
4 **Gérard Besson, Joachim Lohkamp, Pierre Pansu, and Peter Petersen Miroslav Lovric, Maung Min-Oo, and McKenzie Y.-K. Wang, Editors,** Riemannian geometry, 1996
3 **Albrecht Böttcher, Aad Dijksma and Heinz Langer, Michael A. Dritschel and James Rovnyak, and M. A. Kaashoek Peter Lancaster, Editor,** Lectures on operator theory and its applications, 1996
2 **Victor P. Snaith,** Galois module structure, 1994
1 **Stephen Wiggins,** Global dynamics, phase space transport, orbits homoclinic to resonances, and applications, 1993

Ordered Exponential Fields

Fields Institute Monographs

The Fields Institute for Research in Mathematical Sciences

Ordered Exponential Fields

Salma Kuhlmann

American Mathematical Society
Providence, Rhode Island

The Fields Institute
for Research in Mathematical Sciences

The Fields Institute is named in honour of the Canadian mathematician John Charles Fields (1863–1932). Fields was a visionary who received many honours for his scientific work, including election to the Royal Society of Canada in 1909 and to the Royal Society of London in 1913. Among other accomplishments in the service of the international mathematics community, Fields was responsible for establishing the world's most prestigious prize for mathematics research—the Fields Medal.

The Fields Institute for Research in Mathematical Sciences is supported by grants from the Ontario Ministry of Education and Training and the Natural Sciences and Engineering Research Council of Canada. The Institute is sponsored by McMaster University, the University of Toronto, the University of Waterloo, and York University and has affiliated universities in Ontario and across Canada.

This research was supported by a Deutsche Forschungsgemeinshaft Habilitationsstipendium and an Auslandsaufenthalts-Stipendium. Partially supported by an Individual Research Grant from the Natural Sciences and Engineering Research Council of Canada, and by the University of Saskatchewan President's NSERC fund.

1991 *Mathematics Subject Classification.* Primary 03C60, 12J15; Secondary 12L12, 26A12.

ABSTRACT. We provide a detailed valuation theoretic description of ordered fields which admit an exponential function. In particular, we analyze the structure of the non-archimedean models of o-minimal expansions of the reals, in which the exponential function is definable. We apply our results to study the Hardy fields associated to such expansions. The appendix presents the model theory of the value groups of ordered exponential fields.

Library of Congress Cataloging-in-Publication Data

Kuhlmann, Salma, 1958–
  Ordered exponential fields / Salma Kuhlmann.
    p. cm. — (Fields Institute monographs, ISSN 1069-5273 ; 12)
  Includes bibliographical references and index.
  ISBN 0-8218-0943-1 (acid-free paper)
  1. Model theoretic algebra. 2. Ordered fields. I. Title. II. Series.
QA9.7.K84 2000
511'.8–dc21   99-049502

Copying and reprinting. Individual readers of this publication, and nonprofit libraries acting for them, are permitted to make fair use of the material, such as to copy a chapter for use in teaching or research. Permission is granted to quote brief passages from this publication in reviews, provided the customary acknowledgment of the source is given.

Republication, systematic copying, or multiple reproduction of any material in this publication is permitted only under license from the American Mathematical Society. Requests for such permission should be addressed to the Assistant to the Publisher, American Mathematical Society, P. O. Box 6248, Providence, Rhode Island 02940-6248. Requests can also be made by e-mail to reprint-permission@ams.org.

© 2000 by the American Mathematical Society. All rights reserved.
The American Mathematical Society retains all rights
except those granted to the United States Government.
Printed in the United States of America.

∞ The paper used in this book is acid-free and falls within the guidelines
established to ensure permanence and durability.
This publication was prepared by The Fields Institute.
Visit the AMS home page at URL: http://www.ams.org/

10 9 8 7 6 5 4 3 2 1   05 04 03 02 01 00

à mes soeurs
Magda, Nawal
Farida et Amira

et mes filles
Anna Noura et Naila

Il faut faire de la vie
   un rêve
   et du rêve
   une réalité

   Marie Curie

Contents

Introduction	xiii
Chapter 0. Preliminaries on valued and ordered modules	1
1. Valued modules	1
2. Valuation independence	5
3. Ordered modules	7
Chapter 1. Non-archimedean exponential fields	15
1. The natural valuation of an ordered field	15
2. The skeleton of $(K^{>0}, \cdot, 1, <)$	18
3. Formally exponential fields	22
4. Lexicographic (de)composition of exponentials	24
5. Exponentiation in power series fields	27
6. Extensions and maximality	29
7. The structure theory for countable exponential fields	31
Chapter 2. Valuation theoretic interpretation of the growth and Taylor axioms	33
1. The axiom schemes (GA) and (T)	33
2. (GA)-exponentials and the value group	34
3. Lifting exp from the residue field	36
4. (T)-exponentials on the infinitesimals	37
5. Conclusion	39
6. Countable exponential fields with growth properties	40
7. Natural contractions arising from logarithms	44
Chapter 3. The exponential rank	49
1. Convex valuations	49
2. The exponential analogue of the rank	52
3. (GA)- and (T_1)-prelogarithms	53
4. The shift map ζ_ℓ	56
5. Characterization of the exponential and the principal exponential rank	61
Chapter 4. Construction of exponential fields	65
1. w-Logarithmic cross-sections	65
2. A combinatorial result and its consequences	67
3. Existence of logarithmic cross-sections	70
4. From prelogarithms to logarithms	73
Chapter 5. Models for the elementary theory of the reals with restricted analytic functions and exponentiation	77
1. Twisting a group cross-section by an automorphism	77

2.	The exponential-logarithmic power series field	79
3.	Models of arbitrary principal exponential rank	83

Chapter 6. Exponential Hardy fields — 89
1. Some basic valuation theory — 89
2. Hardy fields — 92
3. Value groups — 97
4. The Hardy field of a polynomially bounded + (exp) expansion — 98
5. Exponential boundedness — 101
6. Levels — 102
7. The Crucial Lemma for models of T_{an} — 103
8. Residue fields of $\mathcal{F}$-exp-log-closures — 107
9. A truncation free solution to the Hardy problem — 111
10. Undefinability of the Riemann ζ-function — 113

Appendix A. The model theory of contraction groups — 117
1. Preliminaries — 117
2. Cuts in ordered Abelian groups — 118
3. Ordered abelian groups with contractions — 120
4. Weak o-minimality — 136

Bibliography — 155

Index — 159

List of Notation — 163

Introduction

The aim of this monograph is to describe the models of the elementary theory of an o-minimal expansion of the reals in which the exponential function is definable. We focus on polynomially bounded + (exp) expansions (cf. Section 4 of Chapter 6). Important examples of such expansions are: the expansion by restricted analytic functions and the unrestricted exponential function (cf. [**D–M–M1**]), the expansion by convergent generalized power series and the unrestricted exponential function (in which the Riemann ζ-function restricted to $(1, \infty)$ is definable: cf. [**D–S1**]), and the expansion by multisummable real power series and the unrestricted exponential function (in which the Γ-function restricted to $(0, \infty)$ is definable: cf. [**D–S2**]).

The notion of o-minimality (cf. Section 2 of Chapter 6) was introduced by van den Dries in [**D3**], while studying the expansion $(\mathbb{R}, \exp)$ of the ordered field of the real numbers by the real exponential function. Van den Dries observed that the subsets of the cartesian products $\mathbb{R}^n$ which are parametrically definable in an o-minimal expansion of $\mathbb{R}$ share many of the geometric properties of semi-algebraic sets. For example, a semi-algebraic set has only finitely many connected components, each of them semi-algebraic (cf. [**CO**]). Van den Dries showed that this result remains true if one replaces "semi-algebraic" by "parametrically definable in an o-minimal expansion of $\mathbb{R}$" (cf. also *cell-decomposition* for o-minimal structures [**K–P–S**]). This is a *finiteness theorem*, and van den Dries has set out as a goal to explain the other finiteness phenomena in real algebraic and real analytic geometry as consequences of o-minimality (cf. [**D4**]). The breakthrough was achieved with Wilkie's results on the o-minimality of the reals with exponentiation (cf. [**W1**]). He showed that the expansion of $\mathbb{R}$ by Pfaffian functions *restricted to the closed unit box* (i.e., the functions are set to be identically zero outside the unit box) has a model complete theory. This result may be viewed as a strong refinement of Gabrielov's Theorem (cf. [**GA**]). The latter states that the class of sub-analytic sets is closed under taking complements. Wilkie's Theorem shows that if the restricted analytic functions used to describe a given sub-analytic set A are Pfaffian, then the complement of A may also be described by Pfaffian functions. Wilkie also establishes the model completeness of the elementary theory $T(\exp)$ of $\mathbb{R}$ with the real exponential function exp. This theorem has an important geometric interpretation (cf. [**W1**], p. 1054): call a subset of $\mathbb{R}^n$ *semi-exponential-algebraic* (semi-EA) if it is defined by exponential-polynomial equations and inequalities, and a map from $\mathbb{R}^n$ to $\mathbb{R}^m$ semi-EA if its graph is so, and finally a set to be *sub*-EA if it is the image of a semi-EA set under a semi-EA map. Then the theorem is equivalent to the assertion that the complement of a sub-EA set is a sub-EA set. This, as for the semi-algebraic case, implies that the class of sub-EA sets is also closed under taking closures, interiors and boundaries. Recently (cf. [**W2**]) Wilkie proved a far-reaching generalization, from which it follows that the expansion of the reals by *total* Pfaffian functions is o-minimal as well.

When proving results such as model completeness, the model theorist has to investigate the class of *all* models of a given theory, instead of studying one particular model. At the center of the class of models that we consider lies the ordered field of the reals. But in this class there are also non-archimedean models, which we undertake to describe. Our basic tool for this is valuation theory. Inspired by the theory of real places and real closed fields, and seeking the analogy to the semi-algebraic case, we systematically develop an exponential analogue for all important notions and methods. We use this abstract machinery to describe explicitly the algebraic structure of the models, and to give concrete constructions. These constructions use power series fields (cf. Theorem 5.7 and Theorem 5.11).

The valuation theory that we need is basic. Indeed, we consider ordered fields with convex valuations, whose residue fields are ordered and hence of characteristic zero. There are no deep mysteries in the theory of valued fields in the characteristic zero case. We use, so to say, "shadows" of valuations. In fact, we reduce the valuation theory to the "bones" by going down to the skeletons of the value groups, and yet further down to the value sets of the value groups. This descent makes us deal with those value sets, which are often lexicographic orderings and have some kind of ultrametric structure that is reminiscent of the valuations way up on the fields.

At first sight, this approach may seem too simplistic. However, the power of this method comes from the fact that in most important cases, as in the case of the power series fields, we can lift the whole situation from the value set back up to the field, via the value group.

At the beginning of this work, our approach to the exponential function is equally naive. Although our ultimate aim is to construct extremely well-behaved exponentials (that is, satisfying all the elementary theorems that the real exponential function does), we start by working on very simple exponentials. We just demand that they are order preserving group isomorphisms from the additive group of the field onto the multiplicative group of its positive elements. But even those almost ridiculously weak exponentials impose tremendous restrictions on the structure of the ordered fields that carry them. At the point of this work where we discover that power series fields cannot carry exponentials, we are forced to make our demands on the exponentials, (or more precisely on their compositional inverses, the *logarithms*) even more modest than before. Then, we have to learn to work with prelogarithms, that is non-surjective logarithms. We just have these logarithms that are not even surjective and that enjoy no reasonable properties whatsoever.

Luckily however, we have three important keys to modify and improve those weak prelogarithms, all the way through.

The first is the discovery that power series fields, whilst not carrying surjective logarithms, always carry *prelogarithms*. Thus we develop a standard method to get a surjective logarithm on a countable union of power series fields.

The second key is a result of [**D–M–M1**] who show that power series fields can be naturally made into models of *restricted* real exponentiation, and even more, of all restricted analytic functions. This structure is preserved by taking countable unions of power series fields. Thus we now end up with ordered fields, endowed with an exponential whose *restriction to the unit box* of the field enjoys the same elementary properties as the real exponential restricted to the unit box in the reals.

But we are not done. In fact the exponential thus constructed does *not yet* satisfy the elementary properties of the *unrestricted* real exponential! Here, we use

Introduction

the last key that opens the last door: Ressayre's Theorem [**RE**]. It states that an exponential on an ordered field which satisfies the elementary properties of the restricted real exponential, *and* satisfies the growth axiom scheme (GA), satisfies the elementary properties of the *unrestricted* real exponential as well.

This leads us to a substantial part of the research presented in this monograph. We undertake a systematic study of the growth axiom scheme. Our main idea is to investigate what (GA) imposes on the value group of the field. Descending even further down, we are able to *encode* the growth axiom scheme in the *value set* of the value group of the field. Heuristically, (GA) is encoded in the order automorphisms which the value set carries. It is certainly much easier to understand a totally ordered *set* (the value set) endowed with an automorphism, than to understand a totally ordered *field* endowed with a logarithm. So to say, we get rid of the algebraic structure, but we lose nothing. Indeed, as mentioned already, we are in the privileged situation where we can lift this information up again, from the value set to the ordered field. We develop a canonical method to achieve this lifting. Finally, this allows us to describe the non-archimedean ordered fields endowed with exponentials and restricted analytic functions which satisfy all the axioms of real exponentiation, and which moreover enjoy further interesting properties.

Related to (GA) is the notion of exponential rank. It is a finer measurement of the growth rates of exponentials than just (GA). We encode the exponential rank as well in the automorphisms of the value set. We construct fields of arbitrary exponential ranks, the "exponential-logarithmic power series fields". These canonically defined exponential fields are the exponential analogue of power series fields in the real case. They can carry a multitude of exponentials of *distinct* growth rates, and enjoy further surprising properties of which we provide a detailed account.

Hardy fields provide the most beautiful example of non-archimedean exponential fields. They were introduced by Hardy (cf. [**HD2**]), as "the natural domain for the study of asymptotic analysis". We apply our general structure theorems for exponential fields to this particularly important case. Inspired by Rosenlicht's work on Hardy fields of finite rank (cf. [**RO1**]), we extend the study to the infinite rank case. We give a detailed description of the value groups and residue fields of Exponential Hardy fields (which necessarily have infinite rank). Using our results, we present at the end of the monograph a new proof to a conjecture raised by Hardy concerning the asymptotic behaviour of the "Logarithmico-Exponential" functions. This conjecture was first established in [**D–M–M2**] by different methods.

In Chapter 0, we gather some preliminaries about valuations on *ordered Abelian groups*. We introduce the value set and the skeleton of such a group. These invariants turn out to be very handy throughout this research.

In Chapter 1, we study the necessary conditions that an ordered exponential field K has to satisfy. We show at the end of Chapter 2 that the exponential induces canonically an amazing map (which we call a contraction) on the value group G of K with respect to the natural valuation. A contraction contracts every archimedean class $\neq \{0\}$ to a set $\{a, -a\}$ of two points, and yet maps G *surjectively* onto G in an order preserving way. The class of ordered Abelian groups that are able to carry contractions is much smaller than the class of *all* ordered Abelian groups. It is an elementarily axiomatizable class and has a very well-behaved model theory. This has been worked out in [**KF1**] and [**KF2**], and we present the results of these two papers in the Appendix of this monograph.

At the end of Chapter 1, we study "small" non-archimedean exponential fields. We investigate the following problem: Given an ordered field K, and assuming that its residue field $\overline{K}$ with respect to the natural valuation (which is a subfield of $\mathbb{R}$) is an exponentially closed subfield of the reals, is it possible to lift the real exponential exp to an exponential f of K? We answer this problem for non-archimedean countable fields that are root closed for positive elements. We get a structure theorem for those fields (cf. Theorem 1.44), and show that they admit exponentials f lifting exp from their residue fields if and only if their value group is isomorphic to the lexicographic sum of copies of the additive ordered group $(\overline{K}, +, 0, <)$, taken over the rationals. In Section 6 of Chapter 2, we show that this condition is indeed sufficient to get exponentials satisfying (GA) on the countable field. This theorem provides a method to construct non-archimedean countable exponential fields, given archimedean ones (cf. Example 1.45).

In Chapter 2, we translate the meaning of (GA) and the Taylor axiom scheme (T) into a valuation theoretic language. For example, we show that (T) is equivalent to assertions of the form $v(f(x) - E_n(x)) > v(x^n)$ (where E_n denotes the n-th partial sum in the Taylor expansion of exp). These results are used throughout the later chapters.

In Chapter 3, we introduce prelogarithms and define the exponential rank to be the chain of convex valuation rings which are compatible with the prelogarithm. We characterize the exponential rank through *exponential equivalence* (cf. Section 4 of Chapter 3), as the rank is characterized by the "multiplicative equivalence relation" in the real case. It is worthwhile mentioning here that if K is a model of the elementary theory T of an exponentially bounded expansion of the reals, such that the exponential f is definable, then the valuation rings R_w of valuations w compatible with f are precisely the T-convex valuation rings of K, in the sense of [**D–L**].

So far, we have only described results that are in nice analogy to the theory of real places. But when it comes to existence results, the analogy breaks down. If a field has a place onto an ordered residue field, then the order can be lifted up to the field through the place. It is not surprising that exponentials cannot be lifted through arbitrary places. Indeed, we show in Chapter 4 that power series fields *never* admit exponentials compatible with their canonical valuation. (It is interesting to note that there is an exponential on the surreal numbers, cf. [**G**], but this "power series field" is a proper class.) However, we show that every power series field $\mathbb{R}((G))$ carries a prelogarithm. Indeed, in Section 3 of Chapter 4, we give an explicit formula for the basic prelogarithm $\log_0$ on power series fields with any given value group of the form $\mathbb{R}^{\Gamma_0}$, where Γ_0 is a totally ordered set. Going to the union over an increasing chain of power series fields, we make this logarithm $\log_0$ surjective (cf. Section 4 of Chapter 4). We call the so-obtained field the *exponential-logarithmic power series field* and denote it by $R((\Gamma_0))^{EL}$. The logarithm $\log_0$ does not satisfy (GA), and we develop a method in Chapter 5 to modify $\log_0$. We show that one can use any order preserving map σ on Γ_0 satisfying that $\sigma\gamma > \gamma$ for all γ, to derive from $\log_0$ a logarithm $\log_\sigma$ having the right growth rate. Its inverse $\exp_\sigma$ will then yield a model of real exponentiation (and restricted analytic functions). This method enables us also to construct exponential fields with arbitrary principal exponential ranks. In this way, our construction exhibits the relation between order endomorphisms of the value groups and the growth rates of the constructed exponentials. We also show that $\mathbb{R}((\Gamma_0))^{EL}$ admits countably infinitely many exponentials of distinct exponential rank, for *any* Γ_0. This contrasts the impression of rigidity which is given by the notation $\mathbb{R}((\Gamma_0))^{EL}$ (cf. Example 5.10 and Remark 5.14).

Introduction

Analogous constructions, using power series fields, are given in [**D–M–M2**]. There, a first limit process is employed to obtain a field with non-surjective exponential, and then a second (inverse) limit process renders the exponential surjective. The outcome is a model called the "logarithmic-exponential power series field". In contrast to the construction given in [**D–M–M2**], our construction uses only one limit process. It is an interesting task for future research to compare the models obtained by these two different approaches.

Chapter 6 answers a question raised by Macintyre in a course given at The Fields Institute, during the Algebraic Model Theory Program, November 1996. In [**D–M–M2**], the authors use results of Ressayre and Mourgues to show that Hardy's field LE of Logarithmico-exponential functions admits a truncation-closed embedding in the logarithmic-exponential power series field (see above). Then, they use this particular embedding to prove Hardy's conjecture (cf. Section 9 of Chapter 6 for details) and to show that certain functions, including the Gamma-function and the Riemann ζ-function, cannot be defined using exponential function, logarithm and restricted analytic functions. While lecturing on the results of [**D–M–M2**], Macintyre asked whether their results could be deduced by a "more invariant" version of truncation. Indeed, we derive the results of [**D–M–M2**] without using embeddings in the logarithmic exponential power series field. We replace truncation results by an intrinsic property, which is an assertion about the residue fields of the Hardy fields with respect to arbitrary convex valuations. It is invariant because it does not depend on an embedding in logarithmic exponential power series fields. As a by-product, we get a structure theorem for the Hardy fields associated to a polynomially bounded + (exp) expansion of the reals (cf. Theorem 6.30) and show, amongst other results, that these Hardy fields have levels in the sense of Rosenlicht (cf. [**RO3**]).

Several results presented in this monograph were obtained in the joint papers [**K–K1**], [**K–K2**], [**K–K3**], [**K–K4**] and [**K–K–S1**]. I would like to thank my co-authors Franz-Viktor Kuhlmann and Saharon Shelah for their essential contributions in our joint work. Special thanks are due to Franz-Viktor Kuhlmann for allowing the inclusion of results of [**KF1**] and [**KF2**] as an Appendix, and for proof-reading my manuscript.

I am particularly grateful to Alexander Prestel and Peter Roquette for their constant support during the years I spent in Germany. I am grateful to Sudesh Kaur Khanduja for inviting me to teach a course on exponential fields at her university (Chandigarh, India, 1995), and to Charles Delzell and James Madden for inviting me to give a mini-course on this material during the Special Semester on Real Algebraic Geometry and Ordered Structures (Baton Rouge, 1996).

I would like to thank Lou van den Dries, Askold Khovanskii, Angus Macintyre, David Marker, John Shackell, Patrick Speissegger and Mark Spivakovsky for useful conversations. I am especially endebted to Chris Miller, from whom I have learned a lot in Baton Rouge and at The Fields Institute.

I am endebted to John Martin and the Mathematics Department in Saskatoon for the privilege of teaching an advanced graduate course on this material during the first year of my appointment. I thank the colleagues who attended my course for their constructive comments. Special thanks to Murray Marshall for always being demanding on rigour and clarity, and for proof-reading parts of my manuscript.

I acknowledge the support from the Deutsche Forschungsgemeinschaft, both through a research grant in Germany and a special grant to attend the Program Year on Algebraic Model Theory at The Fields Institute (1996–1997). I thank

Bradd Hart and Matt Valeriote for organizing this exciting year, and The Fields Institute for its hospitality.

Last but not least, I wish to thank my husband and my children for their patience, and for their loving care.

Salma Kuhlmann
Saskatoon, June

CHAPTER 0

Preliminaries on valued and ordered modules

In this chapter, we review some preliminaries that we shall need throughout the monograph. The main notion appearing here is that of the skeleton $S(G)$ of an ordered Abelian group G. The Hahn groups are introduced, and will play a crucial role in the subsequent chapters.

1 Valued modules

All modules considered in this section are left R–modules, for a fixed ring R with 1. The definitions and results of this section also cover the case of valued Abelian groups since they may be considered as $\mathbb{Z}$–modules. If $\{M_i \mid i \in I\}$ is a family of modules, then $\bigoplus_{i \in I} M_i$ will denote the direct sum.

In the sequel, let M be a module and Γ a **chain** (i.e., a totally ordered set). Let ∞ be an element larger than any element of Γ. A surjective map

$$v : M \longrightarrow \Gamma \cup \{\infty\}$$

is a **valuation on** M (and (M, v) is a **valued module**) if for all $x, y \in M$ and $r \in R$, the following holds:
(i) $v(x) = \infty$ if and only if $x = 0$,
(ii) $v(rx) = v(x)$ if $r \neq 0$,
(iii) $v(x - y) \geq \min\{v(x), v(y)\}$.
We call axiom (iii) the **ultrametric inequality**. Axiom (ii) says that the scalar multiplication by nonzero elements preserves the value. So we may speak of a "valued module with value preserving scalar multiplication" in contrast to modules equipped with a map v which only satisfies axioms (i) and (iii). There are also important applications of the latter, more general notion of a valued module. Here, we will only need valued modules satisfying axiom (ii), so we will suppress the specification "with value preserving scalar multiplication". Note that axiom (ii) together with axiom (i) implies that M is torsion free. The following are consequences of the above axioms:
$v(-x) = v(x)$,
$v(x) \neq v(y) \Rightarrow v(x + y) = \min\{v(x), v(y)\}$,
$v(x + y) > v(x) \Rightarrow v(x) = v(y)$.

We call Γ the **value set** of M and write $v(M)$. Note that by this definition,

$$v(M) = \{v(x) \mid 0 \neq x \in M\}.$$

Note that the restriction of v to a submodule of M is a valuation on that submodule.

Let (M_1, v_1), (M_2, v_2) be two valued modules with value sets Γ_1 and Γ_2 respectively. Let
$$h : M_1 \longrightarrow M_2$$
an isomorphism of R–modules. We say that h **preserves the valuation** (or that h is an **isomorphism of valued modules**) if there exists an isomorphism of chains
$$\varphi : \Gamma_1 \longrightarrow \Gamma_2$$
such that for all $x \in M_1$,
$$\varphi(v_1(x)) = v_2(h(x)) \,.$$
We say that (M_1, v_1) and (M_2, v_2) are **isomorphic as valued modules** if such an isomorphism h exists. Similarly, h is an **embedding of valued modules** if h is an isomorphism of valued modules of M_1 onto a submodule of M_2. We will omit "of valued modules" and "as valued modules" if the context is clear. Two valuations v_1 and v_2 on M are called **equivalent** if the identity map on M is an isomorphism between the valued modules (M, v_1) and (M, v_2).

Lemma 0.1 *An isomorphism $h : M_1 \longrightarrow M_2$ preserves the valuation if and only if the map*
$$\tilde{h} : \Gamma_1 \longrightarrow \Gamma_2 \,, \quad \tilde{h}(v_1(x)) = v_2(h(x)) \tag{1}$$
is well-defined and an isomorphism of chains.

By an **ordered system of modules** we mean a pair
$$[\Gamma, \{B(\gamma) ; \gamma \in \Gamma\}]$$
where $\{B(\gamma) \mid \gamma \in \Gamma\}$ is a family of modules indexed by a chain Γ.

Let $S_i = [\Gamma_i, \{B_i(\gamma) \mid \gamma \in \Gamma_i\}]$ be an ordered system of modules, for $i = 1, 2$. We say that S_1 and S_2 are isomorphic if there exists an isomorphism
$$\varphi : \Gamma_1 \longrightarrow \Gamma_2$$
of chains, and for every $\gamma \in \Gamma_1$ an isomorphism
$$\varphi_\gamma : B_1(\gamma) \longrightarrow B_2(\varphi(\gamma))$$
of modules. Then we will call $[\varphi, \{\varphi_\gamma \mid \gamma \in \Gamma_1\}]$ an **isomorphism of ordered systems** and write
$$[\varphi, \{\varphi_\gamma \mid \gamma \in \Gamma_1\}] : S_1 \simeq S_2 \,.$$

Now take a valued module (M, v) with value set $\Gamma = v(M)$. For $\gamma \in \Gamma$, put
$$\begin{aligned} M^\gamma &= \{x \in M \mid v(x) \geq \gamma\} \\ M_\gamma &= \{x \in M \mid v(x) > \gamma\} \,. \end{aligned}$$
Then M^γ, M_γ are submodules satisfying $M_\gamma \subset M^\gamma \subset M$. We put
$$B(M, \gamma) = M^\gamma / M_\gamma.$$
We say that $B(M, \gamma)$ is the **component corresponding to** γ. The **skeleton** of (M, v), denoted by $S(M)$, is the ordered system $[\Gamma, \{B(M, \gamma) \mid \gamma \in \Gamma\}]$. We will write $B(\gamma)$ instead of $B(M, \gamma)$ if the context is clear, and in what follows, $B_i(\gamma)$ instead of $B(M_i, \gamma)$, for $i = 1, 2$.

For every $\gamma \in \Gamma$, the **coefficient map corresponding to** γ is the canonical homomorphism
$$\pi^M(\gamma, -) : M^\gamma \longrightarrow B(\gamma) \text{ defined by } \pi^M(\gamma, x) = x + M_\gamma \,,$$

1. Valued modules

and we write $\pi(\gamma, -)$ instead of $\pi^M(\gamma, -)$ if the context is clear.

The skeleton is an invariant; the following lemma shows that an isomorphism of valued modules induces an isomorphism of their skeletons.

Lemma 0.2 *Suppose that $h : M_1 \longrightarrow M_2$ is an isomorphism of valued modules, and let $\tilde{h}$ be given by (1). Then for all $\gamma \in \Gamma_1$, the map*

$$h_\gamma : B_1(\gamma) \longrightarrow B_2(\tilde{h}(\gamma))$$

defined by

$$\pi^{M_1}(\gamma, x) \mapsto \pi^{M_2}(\tilde{h}(\gamma), h(x))$$

is well-defined and an isomorphism of modules. Hence,

$$[\tilde{h}, \{h_\gamma \mid \gamma \in \Gamma_1\}] : S(M_1) \simeq S(M_2) .$$

The skeleton is a good invariant, i.e., given $[\Gamma, \{B(\gamma) \mid \gamma \in \Gamma\}]$ an ordered system of torsion free modules, it can be realized as the skeleton of a valued module through the following canonical construction. Let $\prod_{\gamma \in \Gamma} B(\gamma)$ the product module. If $s \in \prod_{\gamma \in \Gamma} B(\gamma)$, let the **support** of s be

$$\text{support } s = \{\gamma \mid s(\gamma) \neq 0\} .$$

The direct sum $\bigoplus_{\gamma \in \Gamma} B(\gamma)$ is the submodule of all elements with finite support. We define

$$v_{\min} : \bigoplus_{\gamma \in \Gamma} B(\gamma) \longrightarrow \Gamma \cup \{\infty\}$$

by

$$v_{\min}(s) = \min(\text{support } s) \tag{2}$$

(by convention, $\min \emptyset = \infty$). This is a valuation, and the valued module thus obtained is the **Hahn sum**, denoted by $\coprod_{\gamma \in \Gamma} B(\gamma)$. The **Hahn product**, denoted by $\mathbf{H}_{\gamma \in \Gamma} B(\gamma)$, is the submodule of $\prod_{\gamma \in \Gamma} B(\gamma)$ consisting of all elements with well ordered support, equipped with the valuation $v_{\min}$ which again is defined by (2). We have:

$$S\left(\coprod_{\gamma \in \Gamma} B(\gamma)\right) \simeq [\Gamma, \{B(\gamma) \mid \gamma \in \Gamma\}] \simeq S\left(\mathbf{H}_{\gamma \in \Gamma} B(\gamma)\right) .$$

Suppose that $M_1 \subset M_2$ and $\Gamma_1 \subset \Gamma_2$. We say that (M_2, v_2) is an **extension** of (M_1, v_1) and write $(M_1, v_1) \subset (M_2, v_2)$ if $v_2(x) = v_1(x)$ for all $x \in M_1$. In this case, for every $\gamma \in \Gamma_1$ there exists a natural identification of $B_1(\gamma)$ with a subspace of $B_2(\gamma)$. In this context, if $\gamma \in \Gamma_2 \setminus \Gamma_1$, we set by convention: $(M_1)^\gamma = (M_2)^\gamma \cap M_1$, $(M_1)_\gamma = (M_2)_\gamma \cap M_1$ and $B_1(\gamma) = 0$. If $\Gamma_1 = \Gamma_2$ and for all $\gamma \in \Gamma_1$, $B_1(\gamma) = B_2(\gamma)$, we will say that the extension is **immediate**.

Lemma 0.3 *The extension $(M_1, v_1) \subset (M_2, v_2)$ is immediate if and only if for all non-zero $x \in M_2$ there exists $y \in M_1$ such that $v_2(x - y) > v_2(x)$.*

For example,

$$\coprod_{\gamma \in \Gamma} B(\gamma) \subset \mathbf{H}_{\gamma \in \Gamma} B(\gamma)$$

is immediate.

We say that $M \neq 0$ is **maximally valued** if it does not admit any proper immediate extension.

Let $\{x_i \mid i \in I\} \subset M \setminus \{0\}$ and $M_0 \subset M$ be a submodule. We say that $\{x_i \mid i \in I\}$ is (linearly) **valuation independent** over M_0 if for all $z_0 \in M_0$ and $r_i \in R$ such that $r_i = 0$ for all but finitely many $i \in I$,

$$v\left(\sum_{i \in I} r_i x_i + z_0\right) = \min_{\{i \in I \mid r_i \neq 0\}} \{v(x_i), v(z_0)\}.$$

If this is the case, then in particular, $\{x_i \mid i \in I\}$ is linearly independent over M_0. By convention, $\emptyset$ is valuation independent over M_0. We say that $\{x_i \mid i \in I\}$ is (linearly) **valuation independent** if it is valuation independent over $M_0 = \{0\}$. Note that in the theory of valued fields, there is also the notion of "algebraically valuation independent", but here we will only deal with the above defined notion, so we will omit the specification "linearly".

Using Zorn's Lemma, it may be shown that in every valued module, there exist maximal valuation independent subsets.

If $\{x_i \mid i \in I\} \subset M$, then

$${}_R^M \langle \{x_i \mid i \in I\} \rangle$$

will denote the R-submodule of M generated by the elements x_i (by convention, ${}_R^M \langle \emptyset \rangle = 0$). If the context is clear, we will omit "M" or "R".

Let $M = V$ be a valued vector space with $S(V) = [\Gamma, \{B(\gamma) \mid \gamma \in \Gamma\}]$. For the proofs of the facts that we will now state (Lemma 0.4 to Corollary 0.9), we refer the reader to [GRA1].

Lemma 0.4 *If $\{x_i \mid i \in I\} \subset V$ is maximal valuation independent, then*

$${}_K \langle \{x_i \mid i \in I\} \rangle \subset V$$

is an immediate extension, and

$${}_K \langle \{x_i \mid i \in I\} \rangle \simeq \coprod_{\gamma \in \Gamma} B(\gamma)$$

as valued vector spaces.

A basis $\mathcal{B}$ of V is called a **valuation basis** if it is a valuation independent set. In this case, $\mathcal{B}$ is maximal valuation independent. Since $\coprod_{\gamma \in \Gamma} B(\gamma)$ admits a valuation basis, Lemma 0.4 gives us a characterization of those vector spaces which admit a valuation basis:

Corollary 0.5 *A valued vector space admits a valuation basis if and only if it is isomorphic to the Hahn sum taken over its skeleton.*

The notion of valuation independence is treated in more detail in the following section.

The next theorem is central in the theory of valued vector spaces:

Theorem 0.6 *Suppose that*
(i) V_i and V_i' are valued vector spaces and V_i' is an immediate extension of V_i, for $i = 1, 2$,
(ii) h is an isomorphism of valued vector spaces of V_1 onto V_2,
(iii) V_2' is maximally valued.

2. Valuation independence

Then there exists an embedding h' of valued vector spaces of V_1' in V_2' such that h' extends h. Moreover, h' is an isomorphism of valued vector spaces of V_1' onto V_2' if and only if V_1' is maximally valued.

One consequence is the following theorem which illuminates the role played by the Hahn sums and products:

Theorem 0.7 *Let $\{x_i \mid i \in I\} \subset V$ be maximal valuation independent and h an isomorphism of $_K\langle\{x_i \mid i \in I\}\rangle$ onto $\coprod_{\gamma \in \Gamma} B(\gamma)$. Then there exists an embedding h' of V in $\mathbf{H}_{\gamma \in \Gamma} B(\gamma)$ extending h.*

A sequence of elements $a_\nu \in V$, $\nu < \lambda$ (λ some limit ordinal) is called a **pseudo Cauchy sequence** in (V, v) if $v(a_\rho - a_\sigma) < v(a_\sigma - a_\tau)$ for all ρ, σ, τ with $\rho < \sigma < \tau < \lambda$. It follows from the ultrametric inequality that $v(a_\nu - a_\tau) = v(a_\nu - a_{\nu+1})$ whenever $\nu < \tau < \lambda$. An element a is called a **pseudo limit** of this pseudo Cauchy sequence if $v(a_\nu - a) = v(a_\nu - a_{\nu+1})$ for all $\nu < \lambda$. We say that V is **pseudo complete** if every pseudo Cauchy sequence has a pseudo limit. In [GRA1] it is shown that

Theorem 0.8 *A valued vector space is maximally valued if and only if it is pseudo complete. Further $\mathbf{H}_{\gamma \in \Gamma} B(\gamma)$ is pseudo complete.*

Thus the above theorem yields a characterization of maximally valued vector spaces.

Corollary 0.9 *A valued vector space is maximally valued if and only if it is isomorphic to the Hahn product over its skeleton.*

The above results show how Hahn sums and products provide a constructive and concrete approach to the abstract theory of valued vector spaces. Indeed, up to isomorphism, any valued vector space is sandwiched between the Hahn sum and the Hahn product over its skeleton. Another crucial consequence that we will often use is the following fact: for a fixed skeleton S, there is a valued vector space, determined uniquely up to isomorphism, which admits a valuation basis (respectively, which is maximally valued) and has skeleton S.

The next corollary follows from the foregoing theorems:

Corollary 0.10 *The following assertions are equivalent:*
1) *V is maximally valued and admits a valuation basis*
2) *$\coprod_{\gamma \in \Gamma} B(\gamma) \simeq \mathbf{H}_{\gamma \in \Gamma} B(\gamma)$*
3) *$\coprod_{\gamma \in \Gamma} B(\gamma) = \mathbf{H}_{\gamma \in \Gamma} B(\gamma)$*
4) *every well ordered subset of Γ is finite*
5) *Γ is the inverse of an ordinal.*

2 Valuation independence

Proposition 0.11 *Let $\mathcal{B} \subset V \setminus \{0\}$. Then $\mathcal{B}$ is valuation independent over V_0 if and only if the following holds: for all $n \in \mathbb{N}$ and distinct $b_1, \ldots, b_n \in \mathcal{B}$ with $v(b_1) = \ldots = v(b_n) = \gamma$, the coefficients $\pi^V(\gamma, b_1), \ldots, \pi^V(\gamma, b_n)$ in $B(V, \gamma)$ are linearly independent over $B(V_0, \gamma)$.*

Proof $\Rightarrow$: Suppose there are $b_1, \ldots, b_n \in \mathcal{B}$ with $v(b_1) = \ldots = v(b_n) = \gamma$ such that $\pi^V(\gamma, b_1), \ldots, \pi^V(\gamma, b_n)$ in $B(V, \gamma)$ are not linearly independent over $B(V_0, \gamma)$.

Then there exist nonzero $k_1, \ldots, k_n \in K$ such that $\pi^V(\gamma, \sum k_i b_i) \in B(V_0, \gamma)$. If we choose $z_0 \in V_0$ satisfying $\pi^V(\gamma, z_0) = \pi^V(\gamma, -\sum k_i b_i)$ then we obtain that

$$v(k_1 b_1 + \ldots + k_n b_n + z_0) > \gamma = \min_{\{i \in I | k_i \neq 0\}} \{v(b_i), v(z_0)\},$$

which shows that $\mathcal{B}$ is not valuation independent over V_0.

$\Leftarrow$: Let $\sum_{i \in I} k_i b_i$ be a finite sum of elements $b_i \in \mathcal{B}$ with $k_i \in K$ and let $z_0 \in V_0$. Let $\gamma = \min_{\{i \in I | k_i \neq 0\}} v(b_i)$. If $v(z_0) < \gamma$, then

$$v\left(\sum_i k_i b_i + z_0\right) = v(z_0) = \min_{\{i \in I | k_i \neq 0\}} \{v(b_i), v(z_0)\}.$$

Assume now that $v(z_0) \geq \gamma$. Without loss of generality, let $1, \ldots, n$ be precisely the indices for which $k_i \neq 0$ and $v(b_i) = \gamma$. If $\pi^V(\gamma, b_1), \ldots, \pi^V(\gamma, b_n)$ are linearly independent over $B(V_0, \gamma)$, that is,

$$\pi^V\left(\gamma, \sum_{i \in I} k_i b_i + z_0\right) = \sum_{i=1}^n k_i \pi^V(\gamma, b_i) + \pi^V(\gamma, z_0) \neq 0.$$

This yields

$$v\left(\sum_{i \in I} k_i b_i + z_0\right) = \gamma = \min_{\{i \in I | k_i \neq 0\}} v(b_i) = \min_{\{i \in I | k_i \neq 0\}} \{v(b_i), v(z_0)\}.$$

$\square$

This proposition shows:

Corollary 0.12 *Let $\mathcal{B} \subset V \setminus \{0\}$. Then $\mathcal{B}$ is maximal valuation independent over V_0 if and only if for every $\gamma \in v(V)$,*

$$\mathcal{B}_\gamma = \{\pi^V(\gamma, b) \mid b \in \mathcal{B} \text{ and } v(b) = \gamma\}$$

forms a basis of $B(V, \gamma)$ over $B(V_0, \gamma)$.

A useful consequence of this corollary is the following well known fact.

Lemma 0.13 *Let (V, v) be a valued K-vector space. If W is a finite dimensional subvector space of V having valuation basis $\mathcal{B}$, and if $a \in V$, then $\mathcal{B}$ can be extended to a valuation basis of $W + Ka$.*

Proof The value set of W is just $\{v(b) \mid b \in \mathcal{B}\}$ and thus finite. Hence, there exists some $a_0 \in W$ such that $v(a - a_0) \notin v(W)$ or, if this is not possible, such that $v(a - a_0) \in v(W)$ is maximal. Since the case $a \in W$ is trivial, we may assume $a \notin W$ which yields that $v(a - a_0) \neq \infty$. If $v(a - a_0) \notin v(W)$, then $\mathcal{B} \cup \{a - a_0\}$ is the required valuation basis of $W + Ka$. Now assume that $\gamma := v(a - a_0) \in v(W)$. By the preceding corollary, $\mathcal{B}_\gamma$ forms a basis of $B(W, \gamma)$. If $\pi^V(\gamma, a - a_0)$ would lie in $B(W, \gamma)$ then there would be a linear combination a_1 of the elements in $\mathcal{B}$ with value γ such that $\pi^V(\gamma, a - a_0 - a_1) = 0$. But this would mean that $v(a - a_0 - a_1) > \gamma$, a contradiction to the maximality of γ. This shows that $\pi^V(\gamma, a - a_0)$ cannot be an element of $B(W, \gamma)$. In view of the above proposition, we find that $a - a_0$ is valuation independent over W and again, it follows that $\mathcal{B} \cup \{a - a_0\}$ is the required valuation basis of $W + Ka$. $\square$

By induction we obtain (cf. [BR]):

Corollary 0.14 (Brown) *A valued vector space of countable dimension admits a valuation basis.*

Actually, more is true: If V is countable dimensional over a subspace that admits a valuation basis, then V admits a valuation basis as well (cf. [KS3]). Brown's Theorem implies:

Corollary 0.15 *If two countable dimensional valued vector spaces (over the same field) have isomorphic skeletons, then they are isomorphic.*

Thus the skeleton classifies countable models up to isomorphism. We will use this fact extensively in the context of exponentiation. In fact, Brown's Theorem generalizes to infinitary language. Indeed, the skeleton classifies arbitrary models up to infinitary equivalence; cf. [KS4].

Valuation independent sets may serve to obtain isomorphisms between valued vector spaces, on the basis of the following lemma.

Lemma 0.16 *Let (V, v) and (V', v') be valued K-vector spaces containing (V_0, v) as a common valued subspace and such that $v(V) = v'(V')$. Let $\mathcal{B} \subset V$ and $\mathcal{B}' \subset V'$ be valuation independent over V_0. Suppose that there exists a bijection*

$$h_\mathcal{B} : \mathcal{B} \longrightarrow \mathcal{B}'$$

such that

$$\forall b \in \mathcal{B} : v'(h_\mathcal{B}(b)) = v(b) \ .$$

Then $h_\mathcal{B}$ extends linearly to an isomorphism over V_0

$$h : {}^V \langle \mathcal{B} \cup V_0 \rangle \longrightarrow {}^{V'} \langle \mathcal{B}' \cup V_0 \rangle$$

of valued vector spaces.

Proof We put $W = {}^V \langle \mathcal{B} \cup V_0 \rangle$ and $W' = {}^{V'} \langle \mathcal{B}' \cup V_0 \rangle$. Then $\mathcal{B}$ and $\mathcal{B}'$ are valuation bases of W resp. W' over V_0, and $h : W \to W'$ is given as follows: for every $z \in V_0$ and $\sum_{i \in I} k_i b_i$ a finite linear combination of elements b_i in $\mathcal{B}$,

$$h\left(\sum_{i \in I} k_i b_i + z\right) = \sum_{i \in I} k_i h_\mathcal{B}(b_i) + z \ .$$

Since $\mathcal{B}$ and $\mathcal{B}'$ are valuation bases of W resp. W' over V_0, and since $v'(h_\mathcal{B}(b)) = v(b)$ for every $b \in \mathcal{B}$ by hypothesis, we have

$$\begin{aligned} v'\left(\sum_{i \in I} k_i h_\mathcal{B}(b_i) + z\right) &= \min_{\{i \in I | k_i \neq 0\}} \{v'(h_\mathcal{B}(b_i)), v(z)\} \\ &= \min_{\{i \in I | k_i \neq 0\}} \{v(b_i), v(z)\} \\ &= v\left(\sum_{i \in I} k_i b_i + z\right) \ . \end{aligned}$$

This shows that h preserves the valuation, which in turn yields that h is well-defined and bijective. $\square$

3 Ordered modules

Throughout this monograph, the word "ordered" will mean "totally ordered" for the algebraic structure under consideration.

In this section, let $(R, <_R)$ be a fixed commutative ordered ring with 1. In the sequel, we will write $<$ instead of $<_R$ to relax the notation whenever the context is clear.

Let M be an R–module and $<$ a total order defined on the set M. We say that $(M,<)$ is an ordered $(R,<_R)$–module if its underlying additive group is an ordered Abelian group, that is, if the order $<$ satisfies

1) $\forall x,y \in M : 0 < x$ and $0 < y \implies 0 < x+y$,
2) $x < y$ if and only if $0 < y - x$,

and if the scalar multiplication satisfies

3) $\forall r \in R: 0 < x$ and $0 <_R r \implies 0 < rx$.

Simple consequences are the following rules:

if $x < y$, then $x + z < y + z$ for all $z \in M$,
if $x < y$, then $rx < ry$ for all $r \in R$ with $0 <_R r$,
$0 < x$ if and only if $-x < 0$,
for $0 < x$ we have $r < s$ if and only if $rx < sx$, $\forall r, s \in R$.

Every ordered $(R, <_R)$–module is torsion free. Every ordered Abelian group is an ordered $\mathbb{Z}$–module.

Let $(M_1, <), (M_2, <)$ be ordered $(R, <_R)$–modules and

$$h : M_1 \longrightarrow M_2$$

a homomorphism of modules. We say that h **preserves the order** (or that h is a **homomorphism of ordered modules**) if for all $x, y \in M_1$, $x \leq y$ implies $h(x) \leq h(y)$.

A submodule $N \subset M$ is said to be **convex** (in M) if it satisfies: if $x_1, x_2 \in N$ and $x \in M$ such that $x_1 < x < x_2$, then $x \in N$. The set of all convex submodules of M, ordered by inclusion, is a chain containing $\{0\}$ and M and is closed under unions and intersections. If N is convex in M, the quotient module M/N, equipped with the order induced by

$$x \leq y \implies x + N \leq y + N$$

is an ordered $(R, <_R)$–module, and the canonical homomorphism $M \longrightarrow M/N$ is order preserving.

Note that if $h: M_1 \to M_2$ is a surjective homomorphism of ordered modules with kernel N, then N is convex in M_1 and h induces an isomorphism of ordered modules from M_1/N onto M_2.

For $x \in M$ we define:

$$C_x(M) = \bigcap \{C \mid C \text{ convex submodule and } x \in C\}$$
$$D_x(M) = \bigcup \{D \mid D \text{ convex submodule and } x \notin D\},$$

and by convention,

$$D_0(M) = \{0\}.$$

A convex submodule of M is called **principal** if it is of the form $C_x(M)$ for some $x \in M$. Note that if N is convex in M and $x \in N$, then

$$C_x(N) = C_x(M) \quad \text{and} \quad D_x(N) = D_x(M).$$

The ordered module

$$B_x(M) = C_x(M)/D_x(M)$$

will be called the **component of x in M**. We will write B_x, C_x and D_x if the context is clear. Note that $B_0(M) = \{0\}$.

3. Ordered modules

We put $|x| = \max\{x, -x\}$. For non-zero $x, y \in M$ we will say that x is R–**equivalent to** y and write $x \overset{R}{\sim} y$ if there exists $r \in R$ such that

$$r|x| \geq |y| \quad \text{and} \quad r|y| \geq |x| \,.$$

We say that x is R–**infinitely smaller** than y and write $x \overset{R}{\ll} y$ if $r|x| < |y|$ for all $r \in R$.

We remark the following properties: $\overset{R}{\sim}$ is an equivalence relation, and $\overset{R}{\ll}$ is compatible with this equivalence relation:

$$x \overset{R}{\ll} y \text{ and } x \overset{R}{\sim} z \implies z \overset{R}{\ll} y$$
$$x \overset{R}{\ll} y \text{ and } y \overset{R}{\sim} z \implies x \overset{R}{\ll} z \,.$$

The R–equivalence class of x will be denoted by $[x]_R$, and the set of R–equivalence classes of nonzero elements by Γ. We define an order on Γ in the following way:

$$[y]_R < [x]_R \quad \text{if and only if} \quad x \overset{R}{\ll} y \,.$$

By the above-mentioned properties, this order is well defined and Γ is a chain. We now further assume below that the ring R satisfies the following condition:

$$\forall r \in R \setminus \{0\} \, \exists s \in R : rs \geq 1 \,. \tag{3}$$

Note that this condition is satisfied by every ordered field and by every archimedean ordered ring. From condition (3) on the ring R it follows that for every non-zero $r \in R$ we have that $rx \overset{R}{\sim} x$. Now the proof of the following proposition is straightforward:

Proposition 0.17 *The map*

$$\begin{aligned} v^R : M &\longrightarrow \Gamma \cup \{\infty\} \\ x &\longmapsto [x]_R \end{aligned}$$

is a valuation on M.

We will call v^R the R–natural valuation on M. **Whenever we consider an ordered module as a valued module it will be understood that the valuation is the R–natural valuation, unless otherwise stated.** We will write v instead of v^R if the context is clear. Note that a given map v from M onto some chain, satisfying $v(0) = \infty$, is a valuation and equivalent to the R–natural valuation of M if and only if

$$v(x) > v(y) \iff x \overset{R}{\ll} y \quad \text{for all non-zero } x, y \in M.$$

Let v be any valuation on an ordered module M; v is called **compatible with the order** on M if for all $x, y \in M$ with $x > 0$ and $y > 0$, $v(x) < v(y)$ implies $y < x$. In particular, v^R is compatible with the order on M.

We say that M is R–**archimedean** if $v^R(M)$ contains at most one element. If $x \in M$ with $v^R(x) = \gamma$, then $C_x = M^\gamma$ and $D_x = M_\gamma$; in particular, M is R–archimedean if and only if M does not contain any nontrivial convex submodule. Hence, $B_x = C_x/D_x = B(\gamma)$ is an R–archimedean ordered $(R, <_R)$–module, and the homomorphism $\pi(\gamma, -)$ preserves the order. To indicate the existence of the induced order on the components, we will speak of the **ordered skeleton** of M and write $S(M, <)$, and we will call $B(\gamma)$ the R–**archimedean component**

corresponding to γ. Similarly, we will say that $S(M_1, <)$ and $S(M_2, <)$ are isomorphic, or that $S(M_1)$ and $S(M_2)$ are **isomorphic as ordered skeletons** if there exists an isomorphism

$$\varphi : \Gamma_1 \longrightarrow \Gamma_2$$

of chains, and for every $\gamma \in \Gamma_1$ an isomorphism

$$\varphi_\gamma : B(M_1, \gamma) \longrightarrow B(M_2, \varphi(\gamma))$$

of ordered modules. And so on for all other notions defined in Section 1.

If M and N are ordered modules, then $M \amalg N$ denotes the sum $M \oplus N$ equipped with the lexicographic order. We will not distinguish between external and internal sums, but we will occasionally indicate internal sums by writing "$M' = M \amalg N$" instead of "$M' \simeq M \amalg N$". The following lemma is quite useful. Its proof is left as an exercise.

Lemma 0.18 *a) Let M be an ordered module, C a convex submodule of M and C' a **complement** of C in M, i.e., C' is a submodule of M such that $M = C' \oplus C$. Then $M \simeq C' \amalg C$ as ordered modules.*
b) Let $\eta : M \to N$ a surjective homomorphism of ordered modules and assume that $\ker \eta$ has a complement in M. Then $M \simeq N \amalg \ker \eta$.
c) Let M, N be ordered modules with convex submodules C and D repectively. Assume that $\eta : M \to N$ is an isomorphism of ordered modules such that $\eta(C) = D$. Then the map $\overline{\eta} : M/C \to N/D$ defined by $\overline{\eta}(a + C) = \eta(a) + D$ is a well-defined isomorphism of ordered modules.

For the computation of skeletons, there is an analogue to part b) of the preceding lemma which has the advantage that it does not require the existence of a complement to $\ker \eta$ in M. We first need some notations. If Δ_1, Δ_2 are two chains, then $\Delta_1 + \Delta_2$ will denote their **sum**, that is, the set $(\{1\} \times \Delta_1) \cup (\{2\} \times \Delta_2)$ ordered lexicographically. Now let $S_i = [\Delta_i, \{B_i(\delta) \mid \delta \in \Delta_i\}]$ be ordered systems of modules, for $i = 1, 2$. Then $S_1 \amalg S_2$, called the **sum** of S_1 and S_2, will denote the ordered system

$$[\Delta_1 + \Delta_2, \{A(\gamma) \mid \gamma \in \Delta_1 + \Delta_2\}]$$

where

$$A(\gamma) = \begin{cases} B_1(\delta) & \text{if } \gamma = (1, \delta) \in \{1\} \times \Delta_1 \\ B_2(\delta) & \text{if } \gamma = (2, \delta) \in \{2\} \times \Delta_2 \end{cases}.$$

Lemma 0.19 *Let $\eta : M \to N$ be a surjective homomorphism of ordered modules. Then*

$$S(M) \simeq S(N) \amalg S(\ker \eta) .$$

More precisely, if v and v' are the natural valuations of M and N respectively and the restriction of v to $\ker \eta$ is again denoted by v, then the map

$$w : M \to v'(N) + v(\ker \eta) \cup \{\infty\}$$

given by

$$w(a) = \begin{cases} v'(\eta(a)) & \text{if } a \notin \ker \eta \\ v(a) & \text{if } a \in \ker \eta \end{cases} \tag{4}$$

is well-defined, surjective and equivalent to the natural valuation v on M. Further,

$$\begin{array}{rcll} B_a(M) & \simeq & B_{\eta(a)}(N) & \text{if } a \notin \ker \eta \\ B_a(M) & = & B_a(\ker \eta) & \text{if } a \in \ker \eta. \end{array}$$

3. Ordered modules

Proof Clearly $C \mapsto \eta(C)$ is a bijective correspondence of the convex submodules of M containing $\ker \eta$ to the convex submodules of N, and the convex submodules of M not containing $\ker \eta$ are precisely the proper convex submodules of $\ker \eta$. Under the above correspondence, principal convex submodules correspond. Thus the valuation w defined by (4) is equivalent to the natural valuation v on M.

Since $\ker \eta$ is convex, $B(M, \gamma) = B(\ker \eta, \gamma)$ whenever $\gamma \in v(\ker \eta)$. If $a \notin \ker \eta$, then the minimal convex submodule C_a of M containing a, properly contains $\ker \eta$. The maximal convex submodule D_a of M not containing a, will then also contain $\ker \eta$, and thus we may compute

$$B_a(M) = B(M, v(a)) = C_a/D_a \simeq (C_a/\ker \eta)/(D_a/\ker \eta) \simeq \eta(C_a)/\eta(D_a).$$

By the correspondence that we have described at the beginning of this proof, it follows that $\eta(C_a)$ is the least convex submodule of N containing $\eta(a)$ and that $\eta(D_a)$ is the largest convex submodule of N not containing $\eta(a)$. This shows that $\eta(C_a)/\eta(D_a) = B(N, v(\eta(a))) = B_{\eta(a)}(N)$ which completes our proof. □

Corollary 0.20 *The skeleton of a lexicographic sum $A \amalg B$ is (isomorphic to) the lexicographic sum $S(A) \amalg S(B)$ of their skeletons.*

The following two lemmas will be used in the subsequent chapters. Their proofs rely significantly on the foregoing lemma.

Lemma 0.21 *The lexicographic sum $A \amalg B$ of ordered modules is maximally valued if and only if both A and B are maximally valued.*

Proof $\Rightarrow$: If B is not maximally valued, then there is an immediate extension B' of B, and by virtue of the foregoing lemma, also $A \amalg B'$ will be an immediate extension of $A \amalg B$. A similar argument works if A is not maximally valued.

$\Leftarrow$: Let M be a nontrivial extension of $A \amalg B$. Then the convex hull B' of B in M is an extension of B and the canonical epimorphism $M \to M/B'$ induces an embedding of A in M/B'. If both A and B are maximally valued, then at least one of the extensions $B \subset B'$ and $A \subset M/B'$ is not immediate, and it follows from the foregoing lemma that $A \amalg B \subset M$ cannot be immediate. □

Lemma 0.22 *The lexicographic sum $A \amalg B$ of ordered modules admits a valuation basis if and only if both A and B do.*

Proof Set $M = A \amalg B$ and denote by η the canonical epimorphism onto A with kernel B:

$$\begin{array}{rcl} \eta : M & \longrightarrow & A \\ a + b & \mapsto & a \end{array} \quad \text{for all } a \in A, \, b \in B.$$

We shall use the notation of Lemma 0.19, with A in the place of N. The following fact will be used in the proof: since B is convex in M, we have for every $x \in M$:

$$x \in B \quad \text{if and only if} \quad v(x) \geq v(b) \text{ for some } b \in B. \tag{5}$$

$\Rightarrow$: Let $\{x_i \mid i \in I\}$ be a valuation basis for M and set

$$\mathcal{B} = \{x_i \mid i \in I\} \cap B.$$

If $b \in B$ and $b = \sum_{i \in I} r_i x_i$ (where all but a finite number of the r_i's is zero), then

$$v(b) = \min_{\{i \in I \mid r_i \neq 0\}} v(x_i).$$

Consequently, for all $i \in I$ with $r_i \neq 0$, $v(x_i) \geq v(b)$, so by (5), $x_i \in B$. This shows that $\mathcal{B}$ is a valuation basis of B.

Now let $\mathcal{A} = \eta(\{x_i \mid i \in I\}) \setminus \{0\}$. Clearly, $\mathcal{A}$ is a generating set for A. Moreover, it is valuation independent. Indeed, if $\sum r_i \eta(x_i)$ is a finite linear combination of elements of $\mathcal{A}$, then

$$v'\left(\sum r_i \eta(x_i)\right) = v'\left(\eta(\sum r_i x_i)\right)$$

since η is a homomorphism. On the other hand, in this equation, $\eta(x_i) \neq 0$ (by definition of $\mathcal{A}$), so $x_i \notin B$. Consequently, $v(\sum r_i x_i) = \min_i v(x_i) < v(b)$ for all $b \in B$ (by (5)). It follows (again by (5)) that $\sum r_i x_i \notin B$. Now we can apply Lemma 0.19 to obtain that

$$v'\left(\eta(\sum r_i x_i)\right) = w\left(\sum r_i x_i\right) = \min_i w(x_i) = \min_i v'(\eta(x_i)),$$

which proves the assertion.

$\Leftarrow$: Let $\mathcal{A}$ (resp. $\mathcal{B}$) be a valuation basis of A (resp. of B). Then $\mathcal{A} \cup \mathcal{B}$ is a basis of M. Moreover, if $\sum_{i \in I} r_i a_i$ (resp. $\sum_{j \in J} s_j b_j$) is a finite linear combination with non-zero coefficients of elements of $\mathcal{A}$ (resp. of $\mathcal{B}$), then by Lemma 0.19,

$$w\left(\sum_{i \in I} r_i a_i\right) < w\left(\sum_{j \in J} s_j b_j\right),$$

so

$$w\left(\sum_{i \in I} r_i a_i + \sum_{j \in J} s_j b_j\right) = w\left(\sum_{i \in I} r_i a_i\right) = v'\left(\sum_{i \in I} r_i a_i\right) = \min_i v'(a_i) = \min_i w(a_i)$$

where the last two equalities hold by assumption and Lemma 0.19. It follows that $\mathcal{A} \cup \mathcal{B}$ is valuation independent. $\square$

Let us now consider isomorphisms of ordered modules. If h is such an isomorphism, then it preserves the valuation and induces an isomorphism of the ordered skeletons. But it is the converse to this statement which is important for us:

Proposition 0.23 *Suppose that M_1 and M_2 are ordered modules and that $h : M_1 \longrightarrow M_2$ is an isomorphism of valued modules. If*

$$[\tilde{h}, \{h_\gamma \mid \gamma \in v(M_1)\}] : S(M_1) \simeq S(M_2)$$

is an isomorphism of ordered skeletons, then h preserves the order.

Proof Let $x > 0$ and put $\gamma = v(x)$. Then $\pi^{M_1}(\gamma, x) > 0$ in $B(M_1, \gamma)$, hence $h_\gamma(\pi^{M_1}(\gamma, x)) > 0$. That is, $\pi^{M_2}(v(h(x)), h(x)) = \pi^{M_2}(\tilde{h}(\gamma), h(x)) > 0$ in $B(M_2, \tilde{h}(\gamma))$, whence $h(x) > 0$. $\square$

As a corollary, we obtain the following lemma ([GRA2], Lemma 1). Gravett states it for the case $R = \mathbb{Q}$, but the lemma is true for ordered R as above.

Lemma 0.24 *Suppose that*
(i) M_i and M_i' are ordered R–modules, for $i = 1, 2$,
(ii) $M_i \subset M_i'$ are immediate extensions, for $i = 1, 2$,
(iii) $h' : M_1' \to M_2'$ is an isomorphism of valued R–modules, and $h'|_{M_1} : M_1 \to M_2$ is an isomorphism of ordered R–modules.
Then h' is an isomorphism of ordered R–modules.

3. Ordered modules

Proof Since $h'|_{M_1} : M_1 \to M_2$ preserves the order and $M_i \subset M'_i$ are immediate extensions, we have that $[\tilde{h}', \{h'_\gamma \mid \gamma \in \Gamma_1\}] : S(M'_1) \simeq S(M'_2)$ is an isomorphism of ordered skeletons, hence $h' : M'_1 \to M'_2$ preserves the order by virtue of the foregoing proposition. $\square$

We want to exploit this lemma to obtain the analogues to Corollary 0.5, Theorem 0.6, Corollary 0.9 and Theorem 0.7 in the case of ordered vector spaces. To this end, we need the following definitions. Let $[\Gamma, \{B(\gamma) \mid \gamma \in \Gamma\}]$ be a system of R-archimedean ordered $(R, <_R)$-modules. On $\bigoplus_{\gamma \in \Gamma} B(\gamma)$, we define the lexicographic order $<_l$: for all $s_1, s_2 \in \bigoplus_{\gamma \in \Gamma} B(\gamma)$, $s_1 <_l s_2$ if $s_1(\gamma) < s_2(\gamma)$ for $\gamma = \min \operatorname{support} (s_1 - s_2)$. The ordered $(R, <_R)$-module obtained in this way is called the **lexicographic sum**. Similarly, the **lexicographic product** is the submodule of $\prod_{\gamma \in \Gamma} B(\gamma)$ consisting of all elements with well-ordered support, equipped with the order $<_l$. Then for all s_1, s_2,

$$s_1 \overset{R}{\sim} s_2 \quad \text{if and only if} \quad \min \operatorname{support} s_1 = \min \operatorname{support} s_2 \ .$$

Hence, as a valued $(R, <_R)$-module, the lexicographic sum (resp. the lexicographic product) is isomorphic to the Hahn sum (resp. Hahn product), and we will also denote it by $\coprod_{\gamma \in \Gamma} B(\gamma)$ (resp. $\mathbf{H}_{\gamma \in \Gamma} B(\gamma)$). We have that

$$S\left(\coprod_{\gamma \in \Gamma} B(\gamma)\right) \simeq [\Gamma, \{B(\gamma) \mid \gamma \in \Gamma\}] \simeq S\left(\mathbf{H}_{\gamma \in \Gamma} B(\gamma)\right)$$

as ordered skeletons.

From now on, we will consider an ordered field $(K, <_K)$ and an ordered K–vector space $(V, <)$. In this situation, we have the analogue to Corollary 0.5:

Proposition 0.25 *An ordered vector space admits a valuation basis if and only if it is isomorphic (as ordered vector space) to the lexicographic sum over its ordered skeleton.*

Proof Only "$\Rightarrow$" is nontrivial. Let $\mathcal{B} = \{x_i \mid i \in I\}$ be a valuation basis of $(V, <)$ and let

$$h : V \longrightarrow \coprod_{\gamma \in \Gamma} B(V, \gamma)$$

be the map given by $x_i \mapsto s_i$, where the tuple s_i is defined by

$$s_i(\gamma) = \begin{cases} \pi(\gamma, x_i) & \text{if } \gamma = v(x_i) \\ 0 & \text{otherwise,} \end{cases}$$

and

$$h\left(\sum k_i x_i\right) = \sum k_i s_i \quad \text{for } k_i \in K \ .$$

(h is the isomorphism of valued vector spaces whose existence is stated in Corollary 0.5.) Calculating $\tilde{h}$ and $\{h_\gamma \mid \gamma \in \Gamma\}$, one easily verifies that

$$[\tilde{h}, \{h_\gamma \mid \gamma \in \Gamma\}] : S(V) \simeq S\left(\coprod_{\gamma \in \Gamma} B(V, \gamma)\right)$$

is an isomorphism of ordered skeletons. Hence by Proposition 0.23, h preserves the order. $\square$

A Corollary to Theorem 0.6 and Lemma 0.24 is:

Theorem 0.26 (Hahn Embedding Theorem) *Suppose that*
(i) V_i and V_i' are ordered vector spaces and V_i' is an immediate extension of V_i, for $i = 1, 2$,
(ii) $h : V_1 \to V_2$ is an isomorphism of ordered vector spaces,
(iii) V_2' is maximally valued.
Then there exists an embedding $h' : V_1' \to V_2'$ of ordered vector spaces such that h' extends h. Moreover, h' is an isomorphism of ordered vector spaces if and only if V_1' is maximally valued.

As a corollary, we obtain the analogue to Theorem 0.7:

Theorem 0.27 *Let $\{x_i \mid i \in I\} \subset V$ be a maximal valuation independent subset, and $h :\ _K\langle\{x_i \mid i \in I\}\rangle \to \coprod_{\gamma \in \Gamma} B(V, \gamma)$ an isomorphism of ordered vector spaces. Then there exists an embedding*
$$h' : V \to \mathbf{H}_{\gamma \in \Gamma} B(V, \gamma)$$
of ordered vector spaces, extending h.

As well, we obtain the analogue to Corollary 0.9:

Corollary 0.28 *An ordered vector space is maximally valued if and only if it is isomorphic (as ordered vector space) to the lexicographic product taken over its ordered skeleton.*

CHAPTER 1

Non-archimedean exponential fields

Given an ordered field K, let v be its natural valuation, with value group $G = v(K)$ and residue field $\overline{K}$. Let v_G be the natural valuation of G. Classical results on the model theory of valued fields have successfully shown how to analyze (elementary) properties of a valued field through those of its value group and residue field. One of the aims of this monograph is to achieve analoguous results for exponential fields. In this chapter, we analyze the value group and residue field of an ordered exponential field K. We deduce very strong conditions on these valuation theoretic invariants of K. We prove that these conditions imply the existence of an exponential in case K is countable.

1 The natural valuation of an ordered field

In this section, we will deal with ordered Abelian groups (which are appearing in connection with exponential fields). They may be viewed as ordered $\mathbb{Z}$–modules, and we have the notions of "$\mathbb{Z}$–**equivalent**", "$\mathbb{Z}$–**natural valuation**", "$\mathbb{Z}$–**archimedean**" and "**ordered skeleton**" introduced in Section 3 of Chapter 0. We will abbreviate the terminology by omitting the "$\mathbb{Z}$". Note that the $\mathbb{Z}$–equivalence relation is just the usual archimedean equivalence relation: if G is an ordered Abelian group, $a, b \in G$ are called **archimedean equivalent** if there is some $n \in \mathbb{N}$ such that $n|a| \geq |b|$ and $n|b| \geq |a|$. For example, an ordered Abelian group is "archimedean" if it does not contain proper nontrivial convex subgroups, and we have:

Theorem 1.1 (Hölder) *Every archimedean group is isomorphic (as an ordered group) to a subgroup of* $(\mathbb{R}, +, 0, <)$.

(cf. [FU]). Hence, the skeleton of an ordered Abelian group is an ordered system of subgroups of $\mathbb{R}$.

Let K be a field, G an ordered Abelian group and ∞ an element greater than every element of G. A surjective map

$$w : K \longrightarrow G \cup \{\infty\}$$

is a **valuation** on K if and only if for all $a, b \in K$,
(i) $w(a) = \infty$ if and only if $a = 0$
(ii) $w(ab) = w(a) + w(b)$
(iii) $w(a - b) \geq \min\{w(a), w(b)\}$.
Then we say that (K, w) is a **valued field**. Immediate consequences are:
$w(a) = w(-a)$
$w(a^{-1}) = -w(a)$ for $a \neq 0$
$w(a) \neq w(b) \Rightarrow w(a + b) = \min\{w(a), w(b)\}$.
We write $w(K) = G$ and call it the **value group** of K.

Note that G carries a natural valuation that we shall denote by v_G, and recall from Chapter 0 that v_G is compatible with the order. Thus for all $g, g' \in G^{<0}$,

$$g \leq g' < 0 \implies v_G(g) \leq v_G(g') \tag{6}$$

and

$$v_G(g) < v_G(g') \implies g < g' < 0 \tag{7}$$

for all $g, g' \in G^{<0}$.

The **valuation ring** of w is the ring

$$R_w = \{a \mid a \in K \text{ and } w(a) \geq 0\}$$

and the **valuation ideal** of w is its unique maximal ideal

$$I_w = \{a \mid a \in K \text{ and } w(a) > 0\}.$$

The field R_w/I_w, denoted by Kw, is the **residue field**. For $b \in R_w$, bw will denote its image under the residue map. The **group of units** of the valuation ring is the subgroup

$$\mathcal{U}_w = \{a \mid a \in K \text{ and } w(a) = 0\}$$

of the multiplicative group of R_w, and the **group of 1-units** is the subgroup

$$1 + I_w = \{a \mid w(a-1) > 0\}.$$

An abelian group is **divisible** if it satisfies

$$\forall x \, \exists y : \, ny = x \qquad (0 \neq n \in \mathbb{N}).$$

Let $(K, +, \cdot, 0, 1, <)$ be an ordered field. The set of positive elements of K will be denoted by $K^{>0}$. Then $(K, +, 0, <)$ and $(K^{>0}, \cdot, 1, <)$ are ordered Abelian groups, and $(K, +, 0, <)$ is divisible.

Let v denote the natural valuation on the ordered Abelian group $(K, +, 0, <)$ (obtained as before by mapping $a \in K$, $a \neq 0$ to its **archimedean equivalence class** $[a] := [a]_{\mathbb{Z}}$.) Let $G = v(K)$ denote its value set. Recall that G is totally ordered by

$$[a] < [b] \quad \text{if and only if} \quad b \ll a.$$

On G, we define an addition: $[a] + [b] = [ab]$. Equipped with this addition and order, G becomes an ordered Abelian group with neutral element $[1]$. The natural valuation on $(K, +, 0, <)$

$$\begin{aligned} v : K &\longrightarrow G \cup \{\infty\} \\ a &\mapsto [a] \end{aligned}$$

is a valuation of the field K, which we call the **natural valuation on the ordered field** K. Recall that v is compatible with the order (or is a **convex valuation**), that is,

$$\text{if } a > 0 \text{ and } b > 0, \text{ then } \quad a \leq b \Rightarrow -v(a) \leq -v(b). \tag{8}$$

We set $\text{sign}(0) = 0$ and for $a \in K$, we set $\text{sign}(a) = 1$ if $a > 0$, and $\text{sign}(a) = -1$ if $a < 0$.

From (8), it follows that for all $a, b \in K$:

$$a < b < 0 \implies v(a) \leq v(b) \tag{9}$$
$$v(a) > v(b) \implies |a| < |b|. \tag{10}$$

1. The natural valuation of an ordered field

Similarly,

$$v(a-b) > v(a) \Longrightarrow \operatorname{sign}(a) = \operatorname{sign}(b) \tag{11}$$

and

$$\operatorname{sign}(a) = \operatorname{sign}(b) \Longrightarrow v(a+b) = \min\{v(a), v(b)\}. \tag{12}$$

Further properties of convex valuations are reviewed in Chapter 3. For now let us note that R_v and I_v are convex in K. As ordered (additive) groups, R_v and I_v are just C_1 (the smallest convex subgroup containing 1) and D_1 (the largest convex subgroup not containing 1) respectively. For the natural valuation v, let us denote the field Kv by $\overline{K}$. Hence, $\overline{K}$ equipped with the canonical order is an archimedean ordered field, and it is just the divisible ordered Abelian group C_1/D_1 equipped with the multiplication

$$(a + D_1) \cdot (b + D_1) = ab + D_1 \,.$$

The coefficient map corresponding to $v(1)$

$$\pi(v(1), -) : R_v \longrightarrow \overline{K}$$

is a homomorphism of ordered rings, and it is just the residue map of the valued field (K, v). To abbreviate the notation, we will denote $\pi(v(1), a)$ by $\overline{a}$ and omit "corresponding to $v(1)$". We have:

$$\forall a, b \in R_v : \overline{a} > \overline{b} \Rightarrow a > b \text{ and } a > b \Rightarrow \overline{a} \geq \overline{b}\,. \tag{13}$$

The valuation ideal I_v is called the ideal of **infinitesimals**, and the valuation ring R_v the ring of **finite elements**. The **positive infinite elements** are the elements of $K^{>0} \setminus R_v$, i.e., the elements which are bigger than every element of the subfield $\mathbb{Q}$ of K. We denote $K^{>0} \setminus R_v$ by $\mathbf{P}_K$.

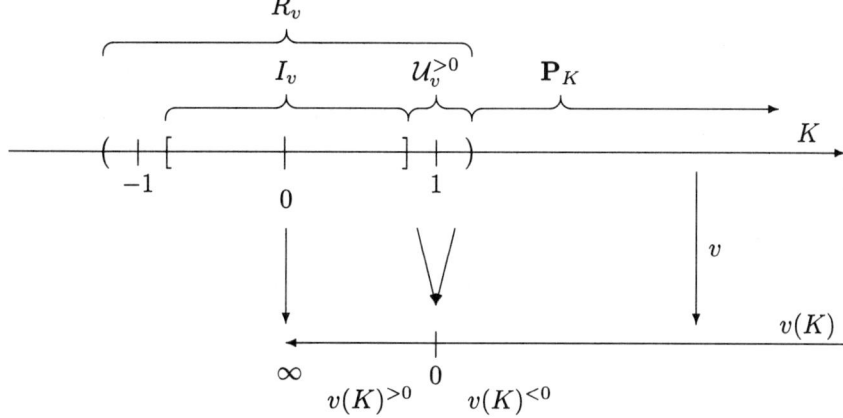

In this monograph, we write v for the natural valuation on an ordered field, v_G for the natural valuation on $G = v(K)$, and $\Gamma = v_G(G)$ for the value set of G. In the sequel we write $G^{<0}$, $G^{>0}$ and $G^{\geq 0}$ for the subsets of negative, positive, and non-negative group elements, respectively.

Since K has multiplication, all archimedean components of $(K, +, 0, <)$ are isomorphic:

Lemma 1.2 *All archimedean components ($\neq 0$) of the divisible ordered Abelian group $(K, +, 0, <)$ are isomorphic to the divisible ordered Abelian group $(\overline{K}, +, 0, <)$.*

Proof Let $a \in K$, $a > 0$. The map
$$C_a \to \overline{K}$$
$$x \mapsto \overline{xa^{-1}}$$
is a surjective homomorphism of ordered groups with kernel D_a. □

Both $\overline{K}$ and $v(K)$ are invariants of the ordered field K:

Lemma 1.3 *An isomorphism of ordered fields preserves the natural valuation and induces isomorphisms of the corresponding residue fields and value groups.*

The divisibility of $(K, +, 0, <)$ enables us to present it as a lexicographic sum of three summands which will play an important role in the course of this monograph. By the convexity of R_v and I_v, Lemmas 0.18, 0.19 and 1.2, we obtain:

Theorem 1.4 (Additive Lexicographic Decomposition) *There exist a group complement $\mathbf{A}$ of R_v in $(K, +, 0, <)$ and a group complement $\mathbf{A}'$ of I_v in R_v such that*
$$(K, +, 0, <) = \mathbf{A} \amalg \mathbf{A}' \amalg I_v . \qquad (14)$$

Both $\mathbf{A}$ and $\mathbf{A}'$ are unique up to order preserving isomorphism, and $\mathbf{A}'$ is order isomorphic to the archimedean group $(\overline{K}, +, 0, <)$. Furthermore, the value set of $\mathbf{A}$ is $G^{<0}$, the one of I_v is $G^{>0}$, and the nonzero components of $\mathbf{A}$ and I_v are all isomorphic to $(\overline{K}, +, 0, <)$.

So we have determined the skeletons $S(\mathbf{A})$ and $S(I_v)$ completely.

2 The skeleton of $(K^{>0}, \cdot, 1, <)$

For the multiplicative group $(K^{>0}, \cdot, 1, <)$ of positive elements, we will now derive a similar decomposition as we have done for the additive group (cf. Theorem 1.8 below). But the multiplicative group is in general not divisible. So, as a hypothesis in Theorem 1.8 we will require the divisibility which actually is equivalent to the property that K is **root closed for positive elements**, that is, for every $a \in K$, $a > 0$, and for every $n \in \mathbb{N}$, there is some $b \in K$ such that $b^n = a$. Note that every real closed field has this property. Note also that in that case, $v(K)$ is divisible.

Let us consider the following subgroups of $\mathcal{U}_v$: the **group of positive units**
$$\mathcal{U}_v^{>0} = \{a \mid a > 0 \text{ and } v(a) = 0\} ,$$
and the **group of 1–units**
$$1 + I_v = \{a \mid v(a - 1) > 0\} .$$

Remark 1.5
(i) By (8), for every $a > 1$,
$$a \in R_v \iff a \in \mathcal{U}_v .$$

(ii) For every $a \in 1 + I_v$ we have $v(a - 1) = v(a^{-1} - 1)$.
(iii) If $a \in \mathcal{U}_v \setminus (1 + I_v)$, then $a^{-1} \in \mathcal{U}_v \setminus (1 + I_v)$.

2. The skeleton of $(K^{>0}, \cdot, 1, <)$

The following two lemmas will give information on the summands of a lexicographic decomposition of $(K^{>0}, \cdot, 1, <)$ (cf. Theorem 1.8) in the case where this group is divisible. But the lemmas are true without this hypothesis and will yield a decomposition of the skeleton of $(K^{>0}, \cdot, 1, <)$ in any case. In view of (8), we have:

Lemma 1.6 *The map*
$$(K^{>0}, \cdot, 1, <) \longrightarrow G$$
$$a \mapsto -v(a) = v(a^{-1})$$
is a surjective homomorphism of ordered groups with kernel $\mathcal{U}_v^{>0}$. It follows that $\mathcal{U}_v^{>0}$ is convex in $(K^{>0}, \cdot, 1, <)$ and that
$$(K^{>0}, \cdot, 1, <)/\mathcal{U}_v^{>0} \simeq G$$
as ordered groups.

Next, we consider $\mathcal{U}_v^{>0}$:

Lemma 1.7 *The map*
$$(\mathcal{U}_v^{>0}, \cdot, 1, <) \longrightarrow (\overline{K}^{>0}, \cdot, 1, <)$$
$$a \mapsto \overline{a}$$
is a surjective homomorphism of ordered groups with kernel $1 + I_v$. It follows that $1 + I_v$ is convex in $(\mathcal{U}_v^{>0}, \cdot, 1, <)$, and
$$(\mathcal{U}_v^{>0}, \cdot, 1, <)/1 + I_v \simeq (\overline{K}^{>0}, \cdot, 1, <) .$$
Consequently, $1 + I_v \subset \mathcal{U}_v^{>0}$ is a jump, i.e., there is no convex subgroup C such that $1 + I_v \subsetneq C \subsetneq \mathcal{U}_v^{>0}$.

Proof Follows immediately from the properties of the residue map together with (8). The last assertion follows from the fact that $\overline{K}$ is archimedean. □

By virtue of Lemmas 0.18, 1.7 and 1.6, we now obtain:

Theorem 1.8 (Multiplicative Lexicographic Decomposition) *If the group $(K^{>0}, \cdot, 1, <)$ is divisible, then there exist a group complement $\mathbf{B}$ of $\mathcal{U}_v^{>0}$ in $(K^{>0}, \cdot, 1, <)$ and a group complement $\mathbf{B}'$ of $1 + I_v$ in $(\mathcal{U}_v^{>0}, \cdot, 1, <)$ such that*
$$(K^{>0}, \cdot, 1, <) = \mathbf{B} \amalg \mathbf{B}' \amalg (1 + I_v, \cdot, 1, <) . \tag{15}$$
Every group complement $\mathbf{B}$ of $\mathcal{U}_v^{>0}$ in $(K^{>0}, \cdot, 1, <)$ is order isomorphic to G through the isomorphism $-v$. Every group complement $\mathbf{B}'$ of $1 + I_v$ in $(\mathcal{U}_v^{>0}, \cdot, 1, <)$ is order isomorphic to $(\overline{K}^{>0}, \cdot, 1, <)$.

In particular $S(\mathbf{B}) = S(G)$. We want to determine $S(1 + I_v)$ as well. So we will compute the natural valuation v and the ordered skeleton of $(K^{>0}, \cdot, 1, <)$ in terms of the natural valuation v_G and the ordered skeleton $[\Gamma, \{B(\gamma) \mid \gamma \in \Gamma\}]$ of G on the one hand, and in terms of the natural valuation v and the ordered skeleton of $(K, +, 0, <)$ on the other hand. Although the first part of the preceding theorem requires the divisibility of the multiplicative group of positive elements of K, we may use Lemmas 1.6 and 1.7 to compute its skeleton even if divisibility does not hold, by means of Lemma 0.19:

Corollary 1.9 *For every ordered field K,*
$$S(K^{>0}, \cdot, 1, <) \simeq S(G) \amalg S(\overline{K}^{>0}, \cdot, 1, <) \amalg S(1 + I_v, \cdot, 1, <) .$$

The value set in $S(\overline{K}^{>0}, \cdot, 1, <)$ consists of just one element since $(\overline{K}^{>0}, \cdot, 1, <)$ is archimedean. Consequently, the only component in this skeleton is itself isomorphic to $(\overline{K}^{>0}, \cdot, 1, <)$.

Our goal is now to give more detailed information on the valuation $v\dot{}$ and the components of $(K^{>0}, \cdot, 1, <)$. For this, we need some notations. As we are working with three different ordered Abelian groups, for "the smallest convex subgroup containing x" and "the biggest convex subgroup not containing x", we write

$$C_x \text{ and } D_x \text{ for } x \in (K, +, 0, <)$$
$$\mathbf{C}_x \text{ and } \mathbf{D}_x \text{ for } x \in (K^{>0}, \cdot, 1, <)$$
$$\mathcal{C}_x \text{ and } \mathcal{D}_x \text{ for } x \in G .$$

In $(K^{>0}, \cdot, 1, <)$, for $\sim, \gg, \ll$ we write $\dot{\sim}, \dot{\gg}$ and $\dot{\ll}$ respectively.

Now we are able to extract additional information from Lemmas 1.6 and 1.7, again by means of Lemma 0.19. Note that for every $a \in K^\times$, we have $v_G(-v(a)) = v_G(v(a))$.

Lemma 1.10 a) Suppose that $a, b \in K^{>0}$ and $v(a) \neq 0$. Then $b \dot{\ll} a$ if and only if $v_G(v(b)) > v_G(v(a))$, and

$$\mathbf{C}_a/\mathbf{D}_a \simeq \mathcal{C}_{v(a)}/\mathcal{D}_{v(a)} .$$

b) Suppose that $a, b \in \mathcal{U}_v^{>0} \setminus 1 + I_v$. Then $a \dot{\sim} b$ and

$$\mathbf{C}_a/\mathbf{D}_a \simeq (\overline{K}^{>0}, \cdot, 1, <).$$

Now it remains to consider $(1 + I_v, \cdot, 1, <)$. We will relate its natural valuation $v\dot{}$ to the natural valuation v of the additive group of K, and we will show that its skeleton is isomorphic to the skeleton of $(I_v, +, 0, <)$.

Lemma 1.11 Suppose that $a > 1$ and $b > 1$. If $v(a-1) = v(b-1)$, then $a \dot{\sim} b$.

Proof We set $\varepsilon_1 = a - 1$ and $\varepsilon_2 = b - 1$. By hypothesis, there exists $n > 1$ such that $n\varepsilon_1 > \varepsilon_2$ and $n\varepsilon_2 > \varepsilon_1$. We write

$$(1+\varepsilon_1)^n = 1 + n\varepsilon_1 + \sum_{i=2}^{n} \binom{n}{i} \varepsilon_1^i .$$

But

$$\varepsilon_1 > 0 \Rightarrow \sum_{i=2}^{n} \binom{n}{i} \varepsilon_1^i > 0 ,$$

hence

$$(1+\varepsilon_1)^n > 1 + n\varepsilon_1 > 1 + \varepsilon_2 .$$

Similarly, one shows that $(1+\varepsilon_2)^n > 1 + \varepsilon_1$. □

Lemma 1.12 Suppose that $a > 1$ and $b > 1$. Let $v(b-1) \geq 0$. If $v(a-1) < v(b-1)$, then $a \dot{\gg} b$.

Proof We set $\varepsilon_1 = a - 1$ and $\varepsilon_2 = b - 1$. We want to show for every $n > 0$: $(1+\varepsilon_2)^n < 1 + \varepsilon_1$. We have that for every $i > 0$,

$$v(\varepsilon_2^i) = iv(\varepsilon_2) \geq v(\varepsilon_2) > v(\varepsilon_1)$$

2. The skeleton of $(K^{>0}, \cdot, 1, <)$

and thus,
$$v\left(\binom{n}{i}\varepsilon_2^i\right) > v(\varepsilon_1),$$
whence
$$v\left(\sum_{i=1}^{n}\binom{n}{i}\varepsilon_2^i\right) > v(\varepsilon_1).$$
In particular (by (8)),
$$\sum_{i=1}^{n}\binom{n}{i}\varepsilon_2^i < \varepsilon_1.$$
On the other hand,
$$(1+\varepsilon_2)^n = 1 + \sum_{i=1}^{n}\binom{n}{i}\varepsilon_2^i,$$
hence,
$$(1+\varepsilon_2)^n < 1 + \varepsilon_1.$$
□

The following corollary will serve us to describe the natural valuation of $1+I_v$:

Corollary 1.13 *If $a > 1$, $b > 1$ and $v(b-1) \geq 0$, then*
$$a \gg b \text{ if and only if } v(a-1) < v(b-1).$$
Consequently, the map
$$a \mapsto v(a-1)$$
is (equivalent to) the natural valuation v on $(1+I_v, \cdot, 1, <)$.

Proof The first assertion is an immediate consequence of Lemmas 1.11 and 1.12. The second assertion is a consequence of the first, by virtue of Remark 1.5 iii). □

Lemma 1.14 *For every $a \in 1+I_v$, the assignment $c \cdot \mathbf{D}_a \mapsto c - 1 + D_{a-1}$ establishes an isomorphism*
$$\mathbf{C}_a/\mathbf{D}_a \simeq (C_{a-1}/D_{a-1}, +, 0, <)$$
of ordered groups.

Proof Assume that $a \in 1+I_v$. We define
$$\phi_a : \mathbf{C}_a \longrightarrow C_{a-1}/D_{a-1}$$
$$c \mapsto c - 1 + D_{a-1}.$$
Let us remark that by definition of $\mathbf{C}_a$ and Corollary 1.13, $c \in \mathbf{C}_a$ if and only if $v(c-1) \geq v(a-1)$. Hence, ϕ_a is well-defined and surjective. Similarly, $c \in \mathbf{D}_a$ if and only if $v(c-1) > v(a-1)$. Hence, $\text{Ker}\,\phi_a = \mathbf{D}_a$.

It remains to show that ϕ_a is a homomorphism. Given $c, d \in \mathbf{C}_a$, we have to prove that $\phi_a(cd) = \phi_a(c) + \phi_a(d)$, i.e., that
$$(cd - 1) - (c - 1) - (d - 1) \in D_{a-1}.$$

This is equivalent to:
$$v((c-1)(d-1)) > v(a-1) .$$

But this is true since
$$v((c-1)(d-1)) = v(c-1) + v(d-1) \geq 2v(a-1) > v(a-1) .$$

Finally, ϕ_a preserves the order: if $c > 1$, then $c - 1 > 0$. □

From Corollary 1.13 and Lemma 1.14, we can deduce the following results:

Theorem 1.15 *For every ordered field K,*
$$S((1+I_v, \cdot, 1, <)) \simeq S((I_v, +, 0, <)) .$$
Moreover if w is given by: $w(a) = v(a-1)$ for $a \in 1 + I_v$, then w is equivalent to the natural valuation $\dot{v}$ on $1 + I_v$.

Proof For v the natural valuation on I_v, Corollary 1.13 shows that w defined by $w(a) = v(a-1)$ is (equivalent to) the natural valuation on $1+I_v$. The component of $1+I_v$ corresponding to $\gamma = w(a)$, $a \in 1+I_v$, is $\mathbf{C}_a/\mathbf{D}_a$ (this follows from Lemma 0.19 in view of the convexity of $1 + I_v$). But by Lemma 1.14, $\mathbf{C}_a/\mathbf{D}_a$ is isomorphic to C_{a-1}/D_{a-1} which is the component of I_v corresponding to $v(a-1) = w(a)$. This proves that the skeletons are isomorphic. □

The main theorem of this section now follows immediately from Corollary 1.13, Lemma 1.14 and Lemma 1.2:

Theorem 1.16 *Define a valuation*
$$w : K^{>0} \longrightarrow \Gamma + G^{\geq 0} + \{\infty\}$$
on the ordered group $(K^{>0}, \cdot, 1, <)$ as follows: $w(1) = \infty$ and for all $a \neq 1$,
$$w(a) = \begin{cases} v_G(v(a)) & \text{if } a \notin \mathcal{U}_v^{>0} \\ 0 & \text{if } a \in \mathcal{U}_v^{>0} \setminus (1 + I_v) \\ v(a-1) & \text{if } a \in (1 + I_v) . \end{cases}$$
Then w is (equivalent to) the natural valuation $\dot{v}$ on $K^{>0}$, i.e.,
$$\forall a, b \in K^{>0} : a \ll b \iff w(a) > w(b) . \tag{16}$$
The component $A(\delta)$ of $(K^{>0}, \cdot, 1, <)$ corresponding to $\delta \in w(K^{>0})$ is given by
$$A(\delta) = \begin{cases} B(\delta) & \text{if } \delta \in \Gamma \\ (\overline{K}^{>0}, \cdot, 1, <) & \text{if } \delta \in G^{\geq 0}, \delta = 0 \\ (\overline{K}, +, 0, <) & \text{if } \delta \in G^{\geq 0}, \delta > 0 . \end{cases}$$

3 Formally exponential fields

Let $(K, +, \cdot, 0, 1, <)$ be an ordered field. We say that K is a **formally exponential field** if there exists
$$f : (K, +, 0, <) \longrightarrow (K^{>0}, \cdot, 1, <)$$
such that f is an isomorphism of ordered groups. A map f with these properties will be called an **exponential** on K. A **logarithm** on K is the compositional inverse $\ell = f^{-1}$ of an exponential. We say (K, f) is an **exponential field** if K is an ordered field and f is an exponential on K. For example, $(\mathbb{R}, \exp)$ is an exponential field, where $\exp$ is the usual exponential function defined on the reals. A main theme of this monograph is to study the non-archimedean extensions of $\mathbb{R}$

3. Formally exponential fields

to which exp extends as well. In the non-archimedean case, the following condition is very useful. We say that v and f are **compatible** or that f is v-**compatible** if further f satisfies that
$$f(R_v) = \mathcal{U}_v^{>0} \quad \text{and} \quad f(I_v) = 1 + I_v .$$
Similarly, ℓ is v-**compatible** if
$$\ell(\mathcal{U}_v^{>0}) = R_v \quad \text{and} \quad \ell(1 + I_v) = I_v . \tag{17}$$

A v-**compatible exponential field** is an exponential field (K, f) such that f is v-compatible. We shall see a generalization of the compatibility notion to any convex valuation w in Chapter 3. There, it will play a crucial role in defining the exponential rank. For now let us just note that v-compatibility is quite natural and useful (see next lemma). Indeed, v-compatible exponentials induce canonically an exponential on the archimedean residue field. Thus, with this condition, we are just "normalizing" the exponential. The outcome is a great simplification in analyzing formally exponential fields (cf. next section). Furthermore, v-compatible exponentials on a formally exponential field are obtained for free, as we observe in Lemma 1.18.

Lemma 1.17 *Let*
$$f : (K, +, 0, <) \to (K^{>0}, \cdot, 1, <)$$
be an exponential on K. Then the following are equivalent:

1) $v(f(1) - 1) = 0$,

2) the map
$$\begin{aligned} \overline{f} : (\overline{K}, +, 0, <) &\to (\overline{K}^{>0}, \cdot, 1, <) \\ \overline{a} &\mapsto \overline{f(a)} \end{aligned}$$
is well-defined and defines an exponential on $\overline{K}$,

3) f is v-compatible.

Proof We first show 1)⇒2). Assume 1) holds. Note that $f(1) \in \mathcal{U}_v^{>0} \setminus 1 + I_v$. Indeed by 1) and the ultrametric inequality, $f(1) \in R_v$ and $f(1) \notin 1 + I_v$. On the other hand, $f(1) > f(0) = 1$, so $f(1)$ is not an infinitesimal. Now, by Lemma 1.2, the archimedean component corresponding to 1 in $(K, +, 0, <)$ is $(\overline{K}, +, 0, <) = R_v/I_v$, and by part b) of Lemma 1.10, that corresponding to $f(1)$ in $(K^{>0}, \cdot, 1, <)$ is $(\mathcal{U}_v^{>0}/1 + I_v, \cdot, 1, <)$. So by Lemma 0.2, the map
$$\begin{aligned} f_0 : \overline{K} &\to \mathcal{U}_v^{>0}/1 + I_v \\ a + I_v &\mapsto f(a) \cdot (1 + I_v) \end{aligned}$$
is an isomorphism of ordered groups. Now note that
$$\mathcal{U}_v^{>0}/1 + I_v = (R_v/I_v)^{>0} = \overline{K}^{>0} .$$
In fact,
$$\forall b \in \mathcal{U}_v^{>0} : b \cdot (1 + I_v) = b + bI_v = b + I_v$$
since $v(b) = 0$ for all $b \in \mathcal{U}_v^{>0}$. So indeed we have that
$$f_0(\overline{a}) = \overline{f(a)} .$$

Assume now that 2) holds. It is immediate to see from the properties of $\overline{f}$ that 3) must hold.

If 3) holds, then for every $a \in R_v$ we have $v(f(a)) = 0$ and thus $v(f(a)-1) \geq 0$. So 3) implies
$$\forall a \in R_v \setminus I_v : v(f(a) - 1) = 0 .$$
Since $1 \in R_v \setminus I_v$, we see that 1) holds as required. □

Lemma 1.18 $(K, +, \cdot, 0, 1, <)$ *admits an exponential e (thus is formally exponential) if and only if it admits a v-compatible exponential f.*

Proof Let $a \in K^{>0}$ such that $e(a) = 2$, and put $f(x) = e(ax)$. □

Remark 1.19 Note that the condition of v-compatibility is not an elementary axiom if we do not add a symbol for the valuation to our language. However, we can replace the condition by the stronger but elementary axiom "$f(1) = 2$". In the place of 2, we could choose any other rational number > 1. As well, we could take two rational numbers $r_2 > r_1 > 1$ and use the axiom "$r_1 < f(1) < r_2$". Finally, by a scheme of axioms of this sort, $f(1)$ may be fixed to any real number, up to addition of an infinitesimal (an element ε of value $v(\varepsilon) > 0$).

Our goal is now to find a valuation theoretic criterion for an ordered field K to be formally exponential. That is, we want to determine when K admits a v-compatible exponential. For this, the Lexicographic Decomposition Theorems will play a crucial role.

4 Lexicographic (de)composition of exponentials

Remark 1.20 1) Assume that K is formally exponential. Then the multiplicative group of K is divisible since it is isomorphic to the additive group of K. Thus fix a decomposition (15), according to Theorem 1.8. Suppose that f is a v-compatible exponential on K. Since $f(R_v) = \mathcal{U}_v^{>0}$, $\mathbf{A} = f^{-1}(\mathbf{B})$ is a group complement of R_v in $(K, +, 0, <)$. Similarly, since $f(I_v) = 1 + I_v$, $\mathbf{A}' = f^{-1}(\mathbf{B}')$ is a group complement of I_v in R_v. With these groups $\mathbf{A}$ and $\mathbf{A}'$, we have a decomposition (14). Denoting the restriction of f to $\mathbf{A}$ by f_L, the restriction to $\mathbf{A}'$ by f_M and the restriction to I_v by f_R, we have obtained isomorphisms

$$\begin{aligned} f_L : \mathbf{A} &\longrightarrow \mathbf{B} \\ f_M : \mathbf{A}' &\longrightarrow \mathbf{B}' \\ f_R : I_v &\longrightarrow 1 + I_v . \end{aligned}$$

In view of the display of the lexicographic sums, this motivates the following definitions, for every ordered field K with decompositions (14) and (15).

2) An isomorphism f_L from a group complement $\mathbf{A}$ of R_v in $(K, +, 0, <)$ onto a group complement $\mathbf{B}$ of $\mathcal{U}_v^{>0}$ in $(K^{>0}, \cdot, 1, <)$ will be called a v-**left exponential** (a v-**left logarithm** is defined similarly). In view of the uniqueness of the group complement $\mathbf{B}$ (which is isomorphic to G through the isomorphism $-v$) a v-left exponential automatically induces an isomorphism from $\mathbf{A}$ onto G. Conversely, every isomorphism between $\mathbf{A}$ and G induces a v-left exponential, or equivalently a v-left logarithm. Thus there is a one-to-one correspondence between v-left exponentials with domain $\mathbf{A}$ on the one hand, and isomorphisms g from $\mathbf{A}$ onto G on the other hand. Similarly, there is a one-to-one correspondence between v-left logarithms with range $\mathbf{A}$ on the one hand, and isomorphisms h from G onto $\mathbf{A}$ on the other hand. More precisely if a v-left logarithm

$$\ell : \mathbf{B} \to \mathbf{A}$$

4. Lexicographic (de)composition of exponentials

is given, then
$$h_\ell := \ell \circ (-v)^{-1}$$
is an isomorphism of ordered groups from G onto $\mathbf{A}$. By abuse of notation, we shall also denote this isomorphism by h_f if $f = \ell^{-1}$. If f is an exponential and $f_L : \mathbf{A} \to \mathbf{B}$ is the v-left exponential corresponding to a given decomposition, then we shall write h_f instead of h_{f_L}. That is
$$h_f := f_L^{-1} \circ (-v)^{-1}.$$
Conversely, every isomorphism
$$h : G \to \mathbf{A}$$
gives rise to a v-left logarithm
$$h \circ -v \ .$$

3) An isomorphism f_M from a group complement $\mathbf{A}'$ of I_v in R_v onto a group complement $\mathbf{B}'$ of $1+I_v$ in $\mathcal{U}_v^{>0}$ will be called a v-**middle exponential**. In view of the uniqueness of the group complements, it automatically induces an isomorphism between $(\overline{K}, +, 0, <)$ and $(\overline{K}^{>0}, \cdot, 1, <)$, i.e., an exponential on $\overline{K}$. Conversely, every exponential e on $\overline{K}$ induces a v-middle exponential $e' : \mathbf{A}' \simeq \mathbf{B}'$. Thus again we have a correspondence between v-middle exponentials on the one hand, and exponentials on the residue field on the other hand.

4) An isomorphism f_R from I_v onto $1+I_v$ will be called a v-**right exponential**.

The following lemma is very useful. Its proof is straightforward.

Lemma 1.21 *Given a v-left exponential f_L, a v middle exponential f_M and a v-right exponential f_R, then $f_L \amalg f_M \amalg f_R$ is a v-compatible exponential on K, where*
$$\forall a \in \mathbf{A} \,\forall a' \in \mathbf{A}' \,\forall \varepsilon \in I_v : (f_L \amalg f_M \amalg f_R)(a + a' + \varepsilon) := f_L(a) \cdot f_M(a') \cdot f_R(\varepsilon) \tag{18}$$

In particular, if e is an exponential on $\overline{K}$ and e' the corresponding middle exponential (for a given decomposition), then $f = f_L \amalg e' \amalg f_R$ is an exponential satisfying $\overline{f} = e$.

If $\overline{f} = e$, we say that f **lifts** e. We call $f_L \amalg f_M \amalg f_R$ the **lexicographic product of the exponentials** f_L, f_M and f_R. We will see throughout this monograph that left, right and middle exponentials are quite independent one from another (for instance, the existence of one of them implies by no means that of any of the other two). The following proposition shows that existence of a left exponential already puts very restrictive conditions on the value group:

Proposition 1.22 *Assume that K admits a v-left exponential f (or a v-left logarithm). Then*
$$S(G) \simeq [G^{<0}, \{B(\gamma) \mid \gamma \in G^{<0}\}]$$
with each $B(\gamma)$ isomorphic to $(\overline{K}, +, 0, <)$.

Proof By 2) of the above remark, we know that G is isomorphic to some complement $\mathbf{A}$ through the isomorphism h_f. Thus h_f induces an isomorphism of the skeletons. Now $S(\mathbf{A})$ was described in Theorem 1.4. $\square$

Corollary 1.23 *Assume that K admits a v-left exponential f. Then*
(i) Its value group G is divisible,
(ii) If K is non-archimedean, then $\Gamma = v_G(G)$ is a dense linear ordering without endpoints.

Proof Since G is isomorphic to a complement of the valuation ring, G is divisible. In particular, if $G \neq \{0\}$ we have that $G^{<0}$ is a dense linear ordering without endpoints. Thus so is Γ (since $\Gamma \simeq G^{<0}$). □

Recall from Lemma 0.1 that the isomorphism $\tilde{h}_f : v_G(G) \to G^{<0}$ **induced by** h_f is defined by
$$\tilde{h}_f(v_G(g)) = v(h_f(g))$$
for any $g \in G$. (By abuse of notation, we denote this isomorphism also by $\tilde{h}_\ell$ if $f^{-1} = \ell$).

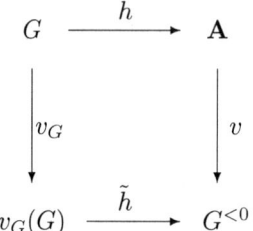

An isomorphism of ordered sets
$$\tilde{h} : v_G(G) \to G^{<0}$$
is called a **group exponential** on G. We say that h is a **lifting** of $\tilde{h}$ if $h : G \to \mathbf{A}$ is an isomorphism such that this diagram commutes.

Thus if K admits a v-left exponential, then G admits a group exponential. Group exponentials will play an essential role throughout this work. Proposition 1.22 motivates the following definition: Let G be an ordered Abelian group with skeleton $[\Gamma, \{B(\gamma) \mid \gamma \in \Gamma\}]$, and A be an archimedean ordered Abelian group. The pair $(G, \tilde{h})$ is an **exponential group in** A if $\tilde{h} : \Gamma \to G^{<0}$ is a group exponential, and $B(\gamma) \simeq A$, for all $\gamma \in \Gamma$.

Note that the group exponential $\tilde{h}_f$ depends only on f and *not* on the chosen lexicographic decompositions:

Proposition 1.24 *Let (K, f) be a v-compatible exponential field and $\ell = f^{-1}$. Assume that $\mathbf{B}_1$ and $\mathbf{B}_1$ are both complements of $\mathcal{U}_v^{>0}$. Let ℓ_1, ℓ_2 be the restrictions of ℓ to $\mathbf{B}_1$ and $\mathbf{B}_2$ respectively. Then $\tilde{h}_{\ell_1} = \tilde{h}_{\ell_2}$.*

Proof Let $g \in G$. By definition
$$\tilde{h}_{\ell_1}(v_G(g)) = v(h_{\ell_1}(g))$$
and
$$\tilde{h}_{\ell_2}(v_G(g)) = v(h_{\ell_2}(g)).$$
Let $b_1 \in \mathbf{B}_1$ and $b_2 \in \mathbf{B}_2$ be such that $-v(b_2) = g = -v(b_1)$. By Theorem 1.16 $v^{\cdot}(b_1) = v^{\cdot}(b_2)$. Hence $b_1 \overset{\cdot}{\sim} b_2$, so $\ell(b_1) \sim \ell(b_2)$. Thus;
$$v(\ell_1(b_1)) = v(\ell(b_1)) = v(\ell(b_2)) = v(\ell_2(b_2)).$$
Now an easy computation ends the proof. □

To summarize, we have the following:

Theorem 1.25 *If (K, f) is a v-compatible exponential field, then $(G, \tilde{h}_f)$ is a divisible exponential group in $(\overline{K}, +, 0, <)$, and $(\overline{K}, \overline{f})$ is an exponential field.*

Note that the first assertion of this theorem already holds under the assumption that f is a v-left exponential on K. This last theorem is just rephrasing a result of Alling ([A], Theorem 1.2 and Corollary 1.4). It describes the necessary conditions that the value group and residue field of a formally exponential field ought to satisfy. A main theme of this monograph is to investigate when the necessary conditions turn out to be sufficient. A complete answer to this question will be presented for power series fields and for countable fields, in the non-archimedean case.

5 Exponentiation in power series fields

Let Γ be any totally ordered set and R any ordered Abelian group. Then R^Γ will denote the Hahn product with index set Γ and components R. (Recall that this is the set of all maps from Γ to R with well-ordered support in Γ). Endowed with the lexicographic order and pointwise addition, R^Γ is an ordered Abelian group. Fix a strictly positive element $1 \in R$ (if R is a field, we take 1 to be the neutral element for multiplication). For every $\gamma \in \Gamma$, we will denote by 1_γ the map which sends γ to 1 and every other element to 0 (1_γ is the characteristic function of the singleton $\{\gamma\}$.)

For G an ordered abelian group, k an ordered field, $k((G))$ will denote the (generalized) **power series field** with coefficients in k and exponents in G. As an ordered Abelian group, this is just the Hahn group k^G. When we work in $K = k((G))$, we will write t^g instead of 1_g. Hence, every element of $k((G))$ can be written in the form $\sum_{g \in G} r_g t^g$ with $r_g \in k$ and well-ordered **support**

$$\{g \in G \mid r_g \neq 0\}.$$

Multiplication is given by the usual formula for multiplying series:

$$\left(\sum_{g \in G} r_g t^g\right)\left(\sum_{g \in G} s_g t^g\right) = \sum_{g \in G} \left(\sum_{g' \in G} r_{g'} s_{g-g'}\right) t^g.$$

$k((G))$ carries a **canonical valuation** given by $v_{\min}(s) = \min(\text{support } s)$ for any series $s \in k((G))$. Clearly the value group is (isomorphic to) G and the residue field is (isomorphic to) k. The valuation ring $k[[G]]$ consists of those series with non-negative exponents, and the valuation ideal of those series with positive exponents. The **constant term** of a series is the coefficient r_0. The units (the series invertible in the valuation ring) are the series with a non-zero constant term. We can truncate a series at its constant term and write it as the sum of two series. Thus a complement of the valuation ring (cf. Theorem 1.4) is the Hahn product $k^{G^{<0}}$. We call it the **canonical additive complement** and denote it by $\text{Neg}(K)$ or by $k((G^{<0}))$. Note that $\text{Neg}(K)$ is in fact a subring.

Similarly, given $s \in K^{>0}$, we can factor out the monomial of smallest exponent $g \in G$ and write $s = t^g u$ with u a series with a positive constant term. Thus a complement of the positive units is the group of (monic) monomials t^g. We call it the **canonical multiplicative complement** and denote it by $\text{Mon}(K)$. Note that $\text{Mon}(K)$ is isomorphic to G through the isomorphism $(-v)(t^g) = -g$ (cf. Theorem 1.8). Note that $v_{\min}$ coincides with the natural valuation if and only if k is archimedean. Power series fields with coefficients in a non-archimedean field k are studied further in Chapter 4.

We need the following notions and results from general valuation theory. An extension of valued fields $(K, w) \subset (L, w)$ is called **immediate** if $w(K) = w(L)$ and $Kw = Lw$. (K, w) is **maximally valued** if it admits no proper immediate extensions. Recall that a valued vector space is maximally valued if and only if every pseudo Cauchy sequence has a pseudo limit ([GRA1]). For the case of valued fields, this was originally shown by Kaplansky [KA]. Power series fields are pseudo complete and thus maximally valued, see [RI].

Lemma 1.26 $(K, w) \subset (L, w)$ *is immediate if and only if for all* $a \in L$, $a \neq 0$ *there exists* $b \in K$ *such that* $w(a - b) > w(a)$.

For a proof, cf. [KF3]. Let (K, w) a valued field and $p(x) = a_n x^n + \ldots + a_0 \in R_w[x]$. The **reduced polynomial** $p(x)w \in Kw[x]$ is the polynomial $p(x)w = (a_n w)x^n + \ldots + a_0 w$, obtained by replacing every coefficient a_i by its residue $a_i w$. A valued field (K, w) is **henselian** if given a polynomial $p(x) \in R_w[x]$, and $a \in Kw$ a simple root of the reduced polynomial $p(x)w \in Kw[x]$, we can find a root $b \in K$ of $p(x)$ such that $bw = a$. Maximal fields are henselian (see [RI]).

Lemma 1.27 *Assume that* (K, w) *is henselian with* char $Kw = 0$. *Then the group of 1-units* $1 + I_w$ *is divisible.*

Proof Let $a = 1 + \varepsilon \in 1 + I_w$ and $n \in \mathbb{N}$. Consider the polynomial $p(x) = x^n - \varepsilon - 1 \in R_w[x]$. Then $p(x)w = x^n - 1$ and 1 is a simple root since char $Kw = 0$. Let $b \in K$ be such that $p(b) = 0$ and $bw = 1$. Thus $b \in 1 + I_w$ and $b^n = a$. □

Let ν, μ, ρ denote ordinals and λ a limit ordinal.

Lemma 1.28 *Let* K *be any ordered field. A sequence* $\{a_\nu \mid \nu < \lambda\}$ *is pseudo Cauchy in* $(1 + I_v, \cdot, 1, <)$ *with respect to* $v^\cdot$ *if and only if* $\{a_\nu - 1 \mid \nu < \lambda\}$ *is pseudo Cauchy in* $(I_v, +, 0, <)$ *with respect to* v. *Moreover,* $a \in 1 + I_v$ *is a pseudo limit of* $\{a_\nu \mid \nu < \lambda\}$ *if and only if* $a - 1 \in I_v$ *is a pseudo limit of* $\{a_\nu - 1 \mid \nu < \lambda\}$.

Proof Let $\rho < \mu < \nu < \lambda$. Then by Corollary 1.13,

$$v^\cdot\left(\frac{a_\nu}{a_\mu}\right) > v^\cdot\left(\frac{a_\mu}{a_\rho}\right) \text{ if and only if } v\left(\frac{a_\nu}{a_\mu} - 1\right) > v\left(\frac{a_\mu}{a_\rho} - 1\right).$$

The latter is equivalent to

$$v\left(\frac{a_\nu - a_\mu}{a_\mu}\right) > v\left(\frac{a_\mu - a_\rho}{a_\rho}\right).$$

Since $v(a_\mu) = v(a_\rho) = 0$, this in turn is equivalent to

$$v(a_\nu - a_\mu) > v(a_\mu - a_\rho),$$

that is,

$$v((a_\nu - 1) - (a_\mu - 1)) > v((a_\mu - 1) - (a_\rho - 1)).$$

This proves the first assertion. The second assertion is proved similarly. □

Corollary 1.29 I_v *is maximally valued with respect to* v *if and only if* $1 + I_v$ *is maximally valued with respect to* $v^\cdot$.

Let E denote any archimedean ordered field and G any ordered Abelian group.

Corollary 1.30 *Every power series field* $E((G))$ *admits a* v-*right exponential.*

Proof Since $E((G))$ is maximally valued, its additive group is maximally valued and consequently, so is $(I_v, +, 0, <)$ (by Lemma 0.21). Hence, it follows from Corollary 1.29 that $(1 + I_v, \cdot, 1, <)$ is maximally valued as well. As a power series field, $E((G))$ is henselian with respect to v. Thus $1 + I_v$ is divisible, that is a $\mathbb{Q}$-vector space. By Theorem 1.15 and Corollary 0.28 it now follows that the ordered groups I_v and $1 + I_v$ must be isomorphic. □

Note that this corollary can also be deduced in a different way using Neumann's Lemma (cf. [N]), as observed by Alling in [A], Section 3, pp. 709-710.

In Chapter 4 (Theorem 4.4), we prove that it is not possible to define left exponentiation on power series fields, thus these fields are never formally exponential.

6 Extensions and maximality

Let (K, f), (L, f') be exponential fields. We say that $(K, f) \subset (L, f')$ is an **extension of exponential fields** if $K \subset L$ is an extension of ordered fields and the restriction of f' to K is f. A v-compatible exponential field (K, f) is **maximal exponential** if it admits no proper immediate (with respect to the natural valuation) v-compatible extension (L, f'). The theory developed so far enables us to characterize the maximal exponential fields. Note that by Theorem 4.4 they cannot be (non-archimedean) power series fields. On the other hand, by [KA] we know that a valued field (K, w) of residue characteristic 0 is a power series field with canonical valuation w if and only if it is maximal. Hence, a (non-archimedean) maximal exponential field is never maximal as a valued field.

We begin by analyzing the extensions of an exponential field which are immediate with respect to the natural valuation. Let $(K, w) \subset (L, w)$ be an extension of valued (not necessarily exponential) fields. Then (K, w) is w-**dense** in (L, w) if for every $a \in L$ and $\alpha \in w(L)$ there is some $b \in K$ such that $w(a - b) > \alpha$. Note that if (L, w) is trivially valued, then (K, w) is w-dense in (L, w) if and only if $K = L$. Note that the valuation w induces a topology on L, called the w-**topology**. A basis for this topology consists of the ultrametric balls (cf. [PC], Chapter 2, Section 4). The notion of w-density introduced here refers to this underlying w-topology.

Lemma 1.31 *If (K, w) is w-dense in (L, w), then the extension is immediate.*

Proof We may assume that (L, w) is not trivially valued. Let $a \neq 0$, $a \in L$. By density there is $b \in K$ such that $w(a - b) > w(a)$. Thus the extension is immediate by Lemma 1.26. □

We need the following characterization of w-density. Assume that $(K, w) \subset (L, w)$, and that K is of characteristic zero (so that $(K, +)$ is divisible). Let $\mathbf{A}_K$ be a group complement of the valuation ring of (K, w). Using Zorn's Lemma, it may be extended to a group complement $\mathbf{A}_L$ for the valuation ring of (L, w).

Lemma 1.32 *Suppose that (L, w) is not trivially valued. Then (K, w) is w-dense in (L, w) if and only if $\mathbf{A}_K = \mathbf{A}_L$.*

Proof If (K, w) is w-dense in (L, w) then for every $a \in L$ there is some $b \in K$ such that $w(a - b) > 0$ which implies that $\mathbf{A}_K$ is already a group complement for the valuation ring of (L, w).

For the converse, assume that the latter holds. Let $a \in L$ and $\alpha \in w(L)$ and choose $0 \neq \alpha' \in w(L)$ such that $\alpha' > \alpha$. We claim that $\alpha' \in w(K)$. Indeed let $d \in L$ be such that $w(d) = -|\alpha'|$. Now $d = a' + c'$ with $a' \neq 0$, $a' \in \mathbf{A}_K$ (so $w(a') < 0$) and $w(c') \geq 0$. By the ultrametric inequality $w(a') = w(d)$ which establishes the claim. Now choose any $c \in K$ with $w(c) = \alpha'$. By hypothesis, for ac^{-1} there exists some $b' \in \mathbf{A}_K$ such that $w(ac^{-1} - b') \geq 0$. Putting $b = b'c$, we find $w(a - b) \geq w(c) = \alpha'$. This proves the converse. □

For ordered fields, there is another notion of density that we would like to review briefly. For more details see [DL–WD]. Let $K \subset L$ be an extension of ordered fields. K is **order dense** in L if for all $a, c \in L$ with $a < c$ there is $b \in K$ such that $a \leq b \leq c$.

In the sequel, let $K \subset L$ be an extension of ordered fields, and denote by v the natural valuation on both of them.

Remark 1.33 1) If (K, v) is v-dense in (L, v), then K is order dense in L.
2) Assume that L is non-archimedean. If K is order dense in L, then (K, v) is v-dense in (L, v).

Corollary 1.34 *Assume that L is non-archimedean. Then K is order dense in L if and only if (K, v) is v-dense in (L, v).*

In fact, if L is non-archimedean, then the interval topology (that is, the topology induced by the order) coincides with the v-topology (cf. [PC]).

Theorem 1.35 *Assume that $(K, f) \subset (L, f)$ is an extension of v-compatible exponential fields and that L is non-archimedean. Let Γ_L, Γ_K denote the value sets of $v(L)$ resp. $v(K)$. Then either (K, v) is v-dense in (L, v), or $\Gamma_L \setminus \Gamma_K$ is infinite.*

Proof Assume that K is archimedean. Then K is not v-dense in L, and the assertion follows since Γ_L is necessarily infinite (Corollary 1.23). Assume now that K is non-archimedean. Let $\mathbf{A}_K \subset \mathbf{A}_L$ as above. If $\mathbf{A}_K = \mathbf{A}_L$, then the foregoing lemma shows that (K, v) is dense in (L, v). Now assume that $\mathbf{A}_K \subset \mathbf{A}_L$ is a proper extension. By Remark 1.20 we then find that also $v(K) \subset v(L)$ is a proper extension and thus, $v(L) \setminus v(K)$ is infinite. Through the isomorphism $\tilde{h}_f$ it follows that $\Gamma_L \setminus \Gamma_K$ is infinite. $\square$

If $\Gamma_L \setminus \Gamma_K$ is infinite, then in particular there are infinitely many values in $v(L)$ that are rationally independent over $v(K)$. Thus there are infinitely many elements in L that are algebraically independent over K. This proves:

Corollary 1.36 *Assume that $(K, f) \subset (L, f)$ is an extension of v-compatible exponential fields and that L is non-archimedean.*
1) If $(K, v) \subset (L, v)$ is immediate, then K is order dense in L.
2) If K is not order dense in L then the transcendence degree of L over K is infinite.

An ordered field K is **order complete** if it admits no proper ordered field extension in which it is order dense. If K is archimedean and order complete, then $K = \mathbb{R}$. Let $K \subset L$ be an extension of ordered fields, L is an **order completion** of K if L is order complete and K is order dense in L. Order completions exist and are unique up to isomorphism. They are maximal with the property that K is order dense in them (cf. [DL–WD]). We denote the order completion of K by K^c. If K is archimedean, then $K^c = \mathbb{R}$. The following corollary follows from Corollary 1.36:

Corollary 1.37 *A non-archimedean v-compatible exponential field which is order complete is maximal exponential.*

In [DL–WD], page 136, it is shown that if f is a v-compatible exponential on K, then by uniform continuity f extends to a v-compatible exponential on K^c. Combining this with the above results we get,

Corollary 1.38 *A non-archimedean v-compatible exponential field (K, f) is maximal exponential if and only if it is order complete.*

7 The structure theory for countable exponential fields

In this section we shall show that the necessary conditions described in Theorem 1.25 are sufficient for countable fields (cf. Corollary 1.41). For the v-right exponentiation on a countable ordered field, we do not even need the necessary conditions:

Corollary 1.39 *Let K be a countable ordered field, root closed for positive elements. Then $I_v \simeq 1 + I_v$ as ordered groups. That is, K admits a v-right exponential.*

Proof Since K is root closed for positive elements, $1 + I_v$ is divisible, thus a $\mathbb{Q}$–vector space. On the other hand, by Corollary 1.15, it has the same skeleton as I_v. So our assertions follow from Brown's Theorem. $\square$

Now the necessary condition on the residue field (in Theorem 1.25) gives us the v-middle exponentiation. Thus it remains to consider v-left exponentiation.

Theorem 1.40 *Let K be a countable ordered field which is root closed for positive elements. Assume that $(v(K), \tilde{h})$ is an exponential group in $(\overline{K}, +, 0, <)$. Then K admits a v-left exponential.*

Proof Set $v(K) = G$. Choose a complement $\mathbf{A}$ of the valuation ring. The hypothesis and the Additive Lexicographic Decomposition Theorem imply that $S(G) \simeq S(\mathbf{A})$. Since both groups are countable $\mathbb{Q}$-vector spaces, the assertion follows by Brown's Theorem. $\square$

Corollary 1.41 *Let K be a countable ordered field which is root closed for positive elements. Assume that $(v(K), \tilde{h})$ is an exponential group in $(\overline{K}, +, 0, <)$, and $(\overline{K}, \overline{f})$ is an exponential field. Then K admits a v-compatible exponential.*

This result is not satisfactory as long as we do not know what the countable exponential groups look like. Fortunately, these groups have a simple structure:

Theorem 1.42 *Let $G \neq 0$ and A be countable divisible ordered Abelian groups and let A be archimedean. Then G is an exponential group in A if and only if $G \simeq \coprod_{\mathbb{Q}} A$.*

Proof $\Rightarrow$: Let G be a countable exponential group in A, denote by Γ its value set. Then $G^{<0} \simeq \Gamma$, so Γ is a countable dense linear ordering without endpoints. By a classical theorem of Cantor (cf. [C]), it follows that $\Gamma \simeq \mathbb{Q}$ as ordered sets. On the other hand, all components of G are isomorphic to A, so by Brown's Theorem we have $G \simeq \coprod_{\mathbb{Q}} A$.
$\Leftarrow$: Let $G \simeq \coprod_{\mathbb{Q}} A$. Then $\Gamma \simeq \mathbb{Q}$, and $G^{<0}$ is a countable dense linear ordering without endpoints. Again by Cantor's Theorem, $G^{<0} \simeq \Gamma$. Since all components of G are A, it follows that G is exponential in A. $\square$

Corollary 1.43 *Let K be a countable non-archimedean ordered field which is root closed for positive elements. Then K admits a v-left exponential if and only if $G \simeq \coprod_{\mathbb{Q}} (\overline{K}, +, 0, <)$.*

We conclude, using the results of this section and Lemma 1.21:

Theorem 1.44 (Countable Case Characterization Theorem) *Let K be a countable non-archimedean ordered field such that $(K^{>0}, \cdot, 1, <)$ is divisible. Given an exponential e on $\overline{K}$, it can be lifted to an exponential f on K (i.e., $\overline{f} = e$) if and only if $G \simeq \coprod_{\mathbb{Q}} (\overline{K}, +, 0, <)$.*

Let E be a subfield of an exponential field (K, f). We say that E is **exponentially closed** in (K, f) if $(E, f|E)$ is an exponential field. In other words, E has to be closed under f and f^{-1}. The last theorem provides a construction method for non-archimedean countable exponential fields, given archimedean ones:

Example 1.45 Let E be a countable exponentially closed and real closed subfield of $(\mathbb{R}, \exp)$ (for example, take E to be the smallest real closed, exp-closed and log-closed subfield containing $\mathbb{Q}$). Set $G \simeq \coprod_{\mathbb{Q}} E$. Consider the power series field $E((G))$. This field is real closed since G is divisible and E is real closed (cf. [PR1]). Consider the subfield F of $E((G))$ generated over E by the monic monomials. That is $F = E(t^g \mid g \in G)$. Clearly F is countable. Let K be the relative algebraic closure of F in $E((G))$. Then K is countable and real closed. Further by our criterion K admits a v-compatible exponential f lifting exp.

CHAPTER 2

Valuation theoretic interpretation of the growth and Taylor axioms

In view of the recent development in the model theory of the real exponential field (cf. [W1]), it is necessary to obtain exponentials satisfying all axioms that are true for the real exponential exp on $\mathbb{R}$. In particular, exp satisfies (infinite schemes of) axioms which relate its growth with the growth of polynomials, and Taylor axioms which express that it is the limit of the sequence of finite partial sums

$$E_n(x) := \sum_{i=0}^{n} \frac{x^i}{i!}.$$

In this chapter we give a valuation theoretic interpretation of the growth and Taylor axioms. We will use this interpretation in Chapter 5 to obtain exponentials satisfying all elementary axioms that are true for exp on $\mathbb{R}$.

1 The axiom schemes (GA) and (T)

For now, we are interested in constructing exponential fields (K, f) satisfying the **growth axiom** scheme:

(GA) $a \geq n^2 \implies f(a) > a^n$ $(n \geq 1)$,

and **Taylor axiom** scheme:

(T) $|a| \leq 1 \implies |f(a) - E_n(a)| < |a^{n+1}|$ $(n \geq 1)$.

We denote by (T_n) the n-th Taylor axiom. A v-compatible exponential is called a **(GA)-exponential** (resp. a **(T)-exponential**, resp. a **(GAT)-exponential**) if it satisfies (GA) (resp. (T), resp. both (GA) and (T)). Further, a v-compatible exponential is called a **(T_n)-exponential** if it satisfies the axioms (T_m) for all $1 \leq m \leq n$, and a **(GAT_n)-exponential** if it is both a (GA)-exponential and a (T_n)-exponential.

We will assume throughout this chapter that every appearing exponential is v-compatible.

Because of the premise "$a \geq n^2$", $f(a) > a^n$ must be verified for all positive infinite elements a, but is void for infinitesimals. Thus it is relevant only in the case of $v(a) \leq 0$. Symmetrically, because of the premise "$|x| \leq 1$", the assertion "$|f(x) - E_n(x)| < |x^{n+1}|$" must be verified for all infinitesimals x, but is void for positive infinite elements. Thus it is relevant only in the case of $v(a) \geq 0$. This establishes:

Proposition 2.1 *f is a (GAT)-exponential on K if and only if the following conditions hold:*
1) *If $a > 0$ and $v(a) < 0$, then $f(a) > a^n$ for all $n \geq 1$,*
2) *If $v(a) = 0$ and $a \geq n^2$ for some $n \geq 1$, then $f(a) > a^n$,*
2') *If $v(a) = 0$ and $|a| \leq 1$, then $|f(a) - E_n(a)| < |a^{n+1}|$ for all $n \geq 1$,*

3) If $v(a) > 0$ then $|f(a) - E_n(a)| < |a^{n+1}|$ for all $n \geq 1$.

Remark 2.2 For future use, it is worthwhile noting a refined version:
(i) f is a (GA)-exponential on K if and only if it satisfies 1) and 2) of the proposition.
(ii) f is a (T)-exponential on K if and only if it satisfies 2') and 3) of the proposition.

A v-left exponential f with domain $\mathbf{A}$ is a **(GA)-v-left exponential** if for all $a > 0$, $a \in \mathbf{A}$ we have that $f(a) > a^n$ for all $n \geq 1$. A v-middle exponential is a **(GAT)-v-middle exponential** if it satisfies (GA) and (T) on its domain. A v-right exponential f is a **(T)-v-right exponential** if $|f(a) - E_n(a)| < |a^{n+1}|$ for all $n \geq 1$ and all $a \in I_v$.

We conclude this section with the following easy but useful observation:

Lemma 2.3 *Let f be a (GAT)-exponential on K. Fix any decomposition (14) of $(K, +, 0, <)$. Let $f_L \amalg f_M \amalg f_R$ be the corresponding decomposition of f. Then*
1) f_L is a (GA)-v-left exponential,
2) f_M a (GAT)-v-middle exponential and
3) f_R is a (T)-v-left exponential.

In the next three sections we analyze those conditions on the left, middle and right exponentials separately.

2 (GA)-exponentials and the value group

Theorem 2.4 (Encoding (GA) in the value group) *Let f be an exponential on K. Fix any decomposition (14) of $(K, +, 0, <)$. Let $f_L : \mathbf{A} \to \mathbf{B}$ be the induced v-left exponential, and $h_f : G \to \mathbf{A}$ the induced isomorphism. Then the following are equivalent:*
1) For all $g \in G^{<0} : v(h_f(g)) > g$,
2) f_L is a (GA)-v-left exponential,
3) If $a > 0$ and $v(a) < 0$, then $f(a) > a^n$ for all $n \geq 1$,
4) If $b \in \mathbf{B}$ and $v(b) < 0$, then $f(b) > b^n$ for all $n \geq 1$.

Proof Set $\ell = f^{-1}$. Assume 4). We show 1) holds. Let $g \in G^{<0}$. Let $b \in \mathbf{B}$ such that $v(b) = g$. Now $v(b) < 0$, so $f(b) > b^n$ for all $n \geq 1$. So $f(b) \gg b$, thus $b \gg \ell(b)$, that is, $g = v(b) < v(\ell(b))$. Now $v(\ell(b)) = v(-\ell(b)) = v(\ell(b^{-1}))$. We compute:
$$h_f(g) = h_f(v(b)) = \ell \circ (-v)^{-1}(v(b)) = \ell(b^{-1}).$$
So 1) holds. Reversing the argument one easily establishes that 1) implies 4).

3) implies 2) and 4) are obvious. We show 2) implies 3). Let $a > 0$ and $v(a) < 0$. Write $a = x + c$ with $x \in \mathbf{A}$, $x > 0$ and $v(c) \geq 0$, (so that $v(x) = v(a)$). By 2) $f_L(x) > x^n$ for all $n \geq 1$. Thus $v(f_L(x)) < v(x^n) = nv(x)$ for all $n \geq 1$. We compute: $v(f(a)) = v(f_L(x) \cdot f(c)) = v(f_L(x)) + v(f(c)) = v(f_L(x)) + 0 < v(x^n) = nv(x) = nv(a) = v(a^n)$ for all $n \geq 1$. Thus $f(a) > a^n$ for all $n \geq 1$ as required.

It remains to show that 4) implies 3). Let $a > 0$ and $v(a) < 0$. Write $a = bc$ with $b \in \mathbf{B}$, and $v(c) = 0$ (so that $v(b) = v(a)$). By 4) $f(b) > b^n$ for all $n \geq 1$. So $f(b) \gg b$, thus $b \gg \ell(b)$, that is $v(b) < v(\ell(b))$. We compute:
$$v(\ell(a)) = v(\ell(b \cdot c)) = v(\ell(b) + \ell(c)).$$

2. (GA)-exponentials and the value group

By v-compatibility, $v(\ell(c)) \geq 0$ (since $v(c) = 0$) and $v(\ell(b)) < 0$ (since $b \in \mathbf{B}$). Thus $v(\ell(b) + \ell(c)) = v(\ell(b))$ (by the ultrametric inequality). Thus,

$$v(\ell(a)) = v(\ell(b)) > v(b) = v(a).$$

Hence $a \gg \ell(a)$, consequently $f(a) \gg a$, that is, $f(a) > a^n$ for all $n \geq 1$. □

The following observation follows immediately by the commutativity of the diagram:

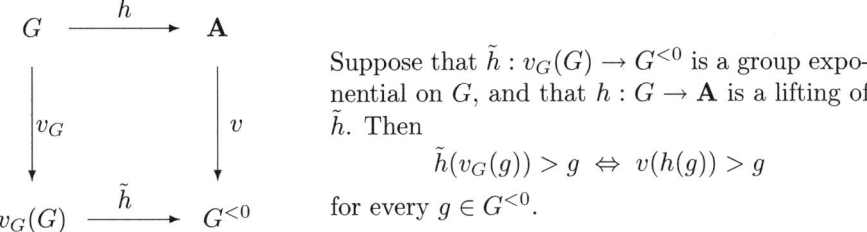

Suppose that $\tilde{h} : v_G(G) \to G^{<0}$ is a group exponential on G, and that $h : G \to \mathbf{A}$ is a lifting of $\tilde{h}$. Then
$$\tilde{h}(v_G(g)) > g \Leftrightarrow v(h(g)) > g$$
for every $g \in G^{<0}$.

We will say that an exponential group $(G, \tilde{h})$ is a **strong exponential group** if for all $g \in G^{<0}$,

$$\tilde{h}(v_G(g)) > g . \tag{19}$$

If this is the case, then $\tilde{h}$ is called a **strong group exponential**. Note that for $\varphi = (\tilde{h})^{-1}$ the condition is equivalent to $\varphi(g) < v_G(g)$ for all $g \in G^{<0}$. Recall that the group exponential $\tilde{h}_f$ depends only on f (cf. Proposition 1.24). We have proved:

Theorem 2.5 *Let f be an exponential on K. Then f satisfies that $f(a) > a^n$ for all $n \geq 1$ (for $a > 0$ and $v(a) < 0$) if and only if $(G, \tilde{h}_f)$ is a strong exponential group.*

Corollary 2.6 (Encoding (GA) in the group exponential) *Let f be an exponential on K. Fix any decomposition (14) of $(K, +, 0, <)$ and let f_L be the induced v-left exponential. Then f_L is a (GA)-v-left exponential if and only if $(G, \tilde{h}_f)$ is a strong exponential group.*

The following theorem gives a condition on the value group G which enables "improving" a v-left exponential to a (GA)-v-left exponential. An ordered Abelian group H has the **lifting property**, if every order automorphism σ of its value set $v_G(H)$ lifts to an order automorphism τ of the group H (lifts means that τ induces σ on $v_G(H)$).

Theorem 2.7 *Assume that K admits an exponential f. Suppose that the value group G has the lifting property, and admits some strong group exponential $\tilde{h}$ (so that $(G, \tilde{h})$ is a strong exponential group). Fix a decomposition (14) of $(K, +, 0, <)$ and let $f_L \amalg f_M \amalg f_R$ be the induced decomposition of f. Then K admits a (GA)-v-left exponential f'_L with the same domain and range as f_L. In particular the exponential $f' = f'_L \amalg f_M \amalg f_R$ satisfies that $f'(a) > a^n$ for all $n \geq 1$ (for $a > 0$ and $v(a) < 0$).*

Proof Assume that $f_L : \mathbf{A} \to \mathbf{B}$. Set $v_G(G) = \Gamma$. Let $h_f : G \to \mathbf{A}$ be the isomorphism induced by f and $\tilde{h}_f : \Gamma \to G^{<0}$ be the group exponential induced by

f (cf. Remark 1.20). Now $\sigma = (\tilde{h}_f)^{-1} \circ \tilde{h}$ is an order automorphism of the value set Γ.

By hypothesis, σ lifts to an automorphism τ of G. Consider the isomorphism $h' : G \to \mathbf{A}$ defined by $h' = h_f \circ \tau$. Let $f'_L : \mathbf{A} \to \mathbf{B}$ be the corresponding v-left exponential (cf. Remark 1.20). Clearly $h_{f'} = h'$. We compute:

$$\begin{aligned}\widetilde{h_{f'}}(v_G(g)) &= v(h'(g)) = v(h_f \circ \tau(g)) = v(h_f(\tau(g))) = \tilde{h}_f(v_G(\tau(g))) \\ &= \tilde{h}_f(\sigma(v_G(g))) = \tilde{h}_f((\tilde{h}_f)^{-1} \circ \tilde{h}(v_G(g))) = \tilde{h}(v_G(g)) \,.\end{aligned}$$

Thus $(G, \tilde{h}_{f'})$ is a strong exponential group. The assertions now follow by Theorems 2.4 and 2.6. □

Note that G has the lifting property if it admits a valuation basis or if it is maximally valued (since the Hahn sum and the Hahn product obviously have the lifting property, cf. Chapter 4). Similarly if $(K, +, 0, <)$ admits a valuation basis, then its value group G has the lifting property. Indeed any complement $\mathbf{A}$ will admit a valuation basis as well (cf. Lemma 0.22). Via the isomorphism h_f, the same is true for G.

Note that also certain intermediate groups between Hahn sum and Hahn product have the lifting property, namely the κ–bounded Hahn products (the subgroups of all elements of a Hahn product whose support has cardinality $< \kappa$). But it does not seem likely that every value group of an exponential field has the lifting property. However, we do not know of an example (which necessarily would have to be uncountable, as we will see in Section 6).

3 Lifting exp from the residue field

We start with the following easy observation:

Lemma 2.8 *Let f be an exponential on a subfield E of $\mathbb{R}$. If f satisfies the axiom scheme*

$$\forall x : |x| < 1 \Rightarrow |f(x) - E_n(x)| \le |x^{n+1}| \quad (\forall n \in \mathbb{N}) \,, \tag{20}$$

then f coincides with the usual exponential exp *and is thus a (GAT)-exponential on E.*

Proof For $|x| < 1$, (20) yields that $f(x)$ is a limit of the series $\sum_{i=0}^{\infty} \frac{x^i}{i!}$. Since E is archimedean, it is the only limit and thus equal to $\exp(x)$.

We have shown that f coincides with exp on the interval $(-1, 1)$. But for every $x \in E$ there is some $n \in \mathbb{N}$ such that $x/n \in (-1, 1)$ (using again that E is archimedean). But f and exp are homomorphisms, thus

$$f(x) = (f(x/n))^n = (\exp(x/n))^n = \exp(x) \,.$$

□

Theorem 2.9 *Let f be an exponential on K. Fix any decomposition (14) of $(K, +, 0, <)$. Let f_M be the induced v-middle exponential, and $\overline{f}$ the induced exponential on $\overline{K}$. Then the following are equivalent:*

1) $\overline{f}$ is a (GAT)-exponential on $\overline{K}$,
2) f_M is a (GAT)-v-middle exponential,
3) f satisfies:
   (i) If $v(a) = 0$ and $a \ge n^2$ for some $n \ge 1$, then $f(a) > a^n$, and
   (ii) If $v(a) = 0$ and $|a| \le 1$, then $|f(a) - E_n(a)| < |a^{n+1}|$ for all $n \ge 1$.

Proof We show that 1) implies 3). The following facts hold in view of (13). If $\overline{f}(\overline{a}) > \overline{a}^n$, then $f(a) > a^n$. If $\overline{f}(\overline{a}) > E_n(\overline{a})$, then $f(a) > E_n(a)$, and if $\overline{f}(\overline{a}) < E_n(\overline{a})$, then $f(a) < E_n(a)$. If $|\overline{f}(\overline{a}) - E_n(\overline{a})| < |\overline{a}^{n+1}|$, then $|f(a) - E_n(a)| < |a^{n+1}|$. This establishes 3). A similar argument works to show that 1) implies 2).

We now show that 3) implies 1). Again, in view of (13), condition 3)(ii) implies that $\overline{f}$ satisfies the axiom scheme

$$\forall x : |x| < 1 \to |\overline{f}(x) - E_n(x)| \le |x^{n+1}| \quad (\forall n \in \mathbb{N}) . \tag{21}$$

Thus by Lemma 2.8 $\overline{f}$ is a (GAT)-exponential as required. A similar argument proves that 2) implies 1). □

4 (T)-exponentials on the infinitesimals

We will consider the following formulas:

$$\begin{aligned} P_n(x,y) &:\equiv |y - E_n(x)| < |x^n| \\ T'_n(x,y) &:\equiv |y - E_n(x)| < |x^{n+1}| \\ Q_n(x,y) &:\equiv v(y - E_n(x)) > v(x^n) . \end{aligned}$$

In the following, we will have a closer look at these formulas and make some basic observations. Note first that $Q_n(a,b)$ trivially holds if $a \ne 0$ and $b = E_n(a)$. From (10), we obtain: for every n, $Q_n(a,b)$ implies $P_n(a,b)$ and if $v(a) < 0$, then $P_n(a,b)$ implies $T'_n(a,b)$. Similarly, $T'_n(a,b)$ implies $P_n(a,b)$ if $v(a) > 0$.

Lemma 2.10 *i)* $Q_n(a,b)$ *implies* $b > E_{n-1}(a)$ *whenever n is even or $a > 0$, and it implies* $b < E_{n-1}(a)$ *whenever n is odd and $a < 0$.*
ii) $Q_n(a,b)$ *implies* $T'_{n-1}(a,b)$, *for every $n > 2$.*
iii) $Q_n(a,b)$ *implies* $Q_m(a,b)$, *whenever $v(a) > 0$ and $m \le n$.*
iv) $T'_n(a,b)$ *implies* $Q_{n-1}(a,b)$, *whenever $v(a) > 0$.*

Proof i): $Q_n(a,b)$ implies $a \ne 0$ and

$$v\left(b - E_{n-1}(a) - \frac{a^n}{n!}\right) > v(a^n) = v\left(\frac{a^n}{n!}\right) , \tag{22}$$

so by (11)

$$\operatorname{sign}(b - E_{n-1}(a)) = \operatorname{sign}\left(\frac{a^n}{n!}\right) .$$

But $a^n/n! > 0$ holds if and only if n is even or a is positive. This proves i).
ii): Equation (22) implies

$$|b - E_{n-1}(a)| = \left|\frac{a^n}{n!} + c\right|$$

for some $c \in a^n I_v$. But then, $v(c) > v(a^n/n!)$ which yields that

$$\left|\frac{a^n}{n!} + c\right| < 2\left|\frac{a^n}{n!}\right| \le |a^n|$$

for $n \ge 2$. This proves assertion ii).
iii): Equation (22) implies

$$v(b - E_{n-1}(a)) = v\left(\frac{a^n}{n!}\right) = v(a^n) . \tag{23}$$

If $v(a) > 0$, then $v(a^n) > v(a^{n-1})$, and thus, equation (23) yields $Q_{n-1}(a,b)$. Our assertion now follows by induction.

iv): Since $|x| < |y|$ implies $v(x) \geq v(y)$, $T'_n(a,b)$ implies $v(b - E_n(a)) \geq v(a^{n+1})$ which in view of $v(a) > 0$ yields equation (22). This in turn yields

$$v(b - E_{n-1}(a)) = v\left(\frac{a^n}{n!}\right) = v(a^n) > v(a^{n-1})$$

which is $Q_{n-1}(a,b)$. □

Corollary 2.11 *Let f be an exponential on K. Then f satisfies $\forall x \in I_v$:*
$$|f(x) - E_n(x)| < |x^{n+1}|$$
for all $n \geq 1$ if and only if $\forall x \in I_v : Q_n(x, f(x))$ holds for all $n \geq 1$.

An important consequence of the last corollary is:

Theorem 2.12 *Let f be an exponential on K. Fix any decomposition (14) of K. Let f_R be the induced v-right exponential. The following assertions are equivalent:*
1) *f satisfies $\forall x \in I_v : Q_n(x, f(x))$ for all $n \geq 1$,*
2) *f_R is a (T)-v-right exponential,*
3) *f satisfies $\forall x$, if $v(x) > 0$ then $|f(x) - E_n(x)| < |x^{n+1}|$ for all $n \geq 1$.*

We need a further lemma about the Q_n–predicates. Note that if K is an ordered field and root closed for positive elements, then for every $y \in K^{>0}$ and $q \in \mathbb{Q}$, the element y^q is a (uniquely defined) element in the divisible (and torsion free) group $(K^{>0}, \cdot, 1, <)$.

Lemma 2.13 *Let K be root closed for positive elements. Suppose that $a \neq 0$, $v(a) > 0$ and $v(c) > 0$. Assume that $Q_n(a,b)$ and $Q_n(c,d)$ hold. Then $Q_n(qa, b^q)$ holds for every $q \in \mathbb{Q}$, $q \neq 0$, and $Q_n(a+c, bd)$ holds if $v(a+c) = \min(v(a), v(c))$.*

Proof From the theory of binomial coefficients, it is known that

$$\left(\sum_{i=0}^n \frac{a^i}{i!}\right) \cdot \left(\sum_{i=0}^n \frac{c^i}{i!}\right) = \sum_{i=0}^n \frac{(a+c)^i}{i!} + A$$

where A is a sum consisting only of monomials $c_{\mu,\nu} a^\mu c^\nu$ for which $\mu + \nu \geq n+1$. W.l.o.g., let $v(a) \leq v(c)$. Then $v(A) \geq v(a^{n+1})$, and we obtain

$$\left(\sum_{i=0}^n \frac{a^i}{i!}\right) \cdot \left(\sum_{i=0}^n \frac{c^i}{i!}\right) \equiv \sum_{i=0}^n \frac{(a+c)^i}{i!} \mod a^{n+1} R_v . \tag{24}$$

Suppose now that $Q_n(a,b)$ and $Q_n(c,d)$ hold. Since both sums on the left side of (24) have v–value 0 and since $v(a) > 0$ and $a \neq 0$, it follows that

$$bd \equiv \sum_{i=0}^n \frac{(a+c)^i}{i!} \mod a^n I_v .$$

Since $v(a+c) = \min(v(a), v(c))$ by the assumption of our lemma, we have $v(a) = v(a+c)$, and the above equivalence is thus nothing else than $Q_n(a+c, bd)$.

Let $r, s > 0$ be arbitrary natural numbers. Replacing both a and c by $\frac{1}{s}a$ we obtain from (24) by induction up to s:

$$\left(\sum_{i=0}^n \frac{(\frac{r}{s}a)^i}{i!}\right)^s \equiv \left(\sum_{i=0}^n \frac{(ra)^i}{i!}\right) \equiv \left(\sum_{i=0}^n \frac{a^i}{i!}\right)^r \mod a^{n+1} R_v .$$

Since the sums have v–value 0, it follows that

$$\left(\sum_{i=0}^{n}\frac{(\frac{r}{s}a)^{i}}{i!}\right) \equiv \left(\sum_{i=0}^{n}\frac{a^{i}}{i!}\right)^{\frac{r}{s}} \mod a^{n+1}R_{v}$$

and, by virtue of $Q_n(a,b)$ and $v(a) > 0$, $a \neq 0$,

$$\left(\sum_{i=0}^{n}\frac{(\frac{r}{s}a)^{i}}{i!}\right) \equiv b^{\frac{r}{s}} \mod a^{n}I_{v}.$$

This proves $Q_n(qa, b^q)$ for all positive rationals q. Finally, it remains to show that $Q_n(a,b)$ implies $Q_n(-a, b^{-1})$. From (24), for $c = -a$ we obtain that

$$\left(\sum_{i=0}^{n}\frac{a^{i}}{i!}\right) \cdot \left(\sum_{i=0}^{n}\frac{(-a)^{i}}{i!}\right) \equiv 1 \mod a^{n+1}R_{v}.$$

Again, since the sums have v–value 0, this yields

$$\left(\sum_{i=0}^{n}\frac{(-a)^{i}}{i!}\right) \equiv \left(\sum_{i=0}^{n}\frac{a^{i}}{i!}\right)^{-1} \mod a^{n+1}R_{v}$$

and, by virtue of $Q_n(a,b)$ and $v(a) > 0$, $a \neq 0$,

$$\left(\sum_{i=0}^{n}\frac{(-a)^{i}}{i!}\right) \equiv b^{-1} \mod a^{n}I_{v}.$$

This proves $Q_n(-a, b^{-1})$ and consequently, $Q_n(qa, b^q)$ for all rationals $q \neq 0$. □

5 Conclusion

From Theorems 2.4, 2.9, 2.12 and Proposition 2.1 we obtain the converse to Lemma 2.3:

Lemma 2.14 *Let f be an exponential on K. Fix any decomposition (14) of $(K, +, 0, <)$. and let $f_L \amalg f_M \amalg f_R$ be the corresponding decomposition of f. If*
1) f_L is a (GA)-v-left exponential,
2) f_M a (GAT)-v-middle exponential and
3) f_R is a (T)-v-right exponential,
then f is a (GAT)-exponential.

This lemma shows how to put together left, middle and right exponentials in order to obtain an exponential with growth properties. Note that the following refined version holds: If f_L is a (GA)-v-left exponential, f_M a (GAT)-v-middle exponential and f_R is a (T_n)-v-right exponential, then f is a (GAT_n)-exponential.

Moreover, using Theorem 2.6 we obtain the following useful valuation theoretic criterion for (GAT)-exponentials. We will use this result in the next section to construct (GAT)-exponentials on countable fields.

Theorem 2.15 *Let f be an exponential on K. Then f is a (GAT)-exponential if and only if*
1) $(G, \tilde{h}_f)$ is a strong exponential group,
2) $\overline{f}$ is a (GAT)-exponential,
3) f satisfies $\forall x \in I_v : Q_n(x, f(x))$ for all $n \geq 1$.

6 Countable exponential fields with growth properties

Recall from Chapter 1 that a nontrivial countable divisible ordered Abelian group G is an exponential group in A (where A is a countable divisible archimedean ordered Abelian group) if and only if $G \simeq \coprod_{\mathbb{Q}} A$. The following proposition will enable us to produce (GA)-exponentials on countable fields.

Proposition 2.16 *Suppose that $A \neq 0$ is any countable divisible archimedean ordered Abelian group, then $\coprod_{\mathbb{Q}} A$ admits a strong group exponential. Hence, every nontrivial countable divisible exponential group is a strong exponential group.*

Proof We shall in fact show more: Suppose that $\{A(q) \mid q \in \mathbb{Q}\}$ is a family of nontrivial countable dense archimedean ordered Abelian groups and set $G = \coprod_{q \in \mathbb{Q}} A(q)$ (hence also G is dense). Then we will show the existence of an isomorphism

$$\varphi : G^{<0} \longrightarrow \mathbb{Q}$$

such that $\forall g : \varphi(g) < v_G(g)$. Then $\tilde{h} = \varphi^{-1}$ will be the required strong group exponential.

Let Φ be the family of all maps ϕ which are order preserving isomorphisms of a finite subset of $G^{<0}$ onto a finite subset of $\mathbb{Q}$ such that $\forall g : \phi(g) < v_G(g)$.

We show that Φ is a nonempty family of partial isomorphisms with the back and forth property (cf. [PO]). Once we have shown that, φ is obtained by a back and forth argument, using induction on countable enumerations of $G^{<0}$ and $\mathbb{Q}$. We use the fact that $\mathbb{Q}$ is dense and without endpoints.

Φ is nonempty: let $g \in G^{<0}$ and $q \in \mathbb{Q}$ such that $q < v_G(g)$. Set $\phi(g) = q$, then $\phi \in \Phi$. Now let $\phi \in \Phi$ and $\dom \phi = \{g_0, \ldots, g_n\}$ with $g_0 < \ldots < g_n$, and $\range \phi = \{q_0, \ldots, q_n\}$ with $q_0 < \ldots < q_n$, and $\phi(g_i) = q_i$ for $0 \leq i \leq n$.

Φ has the back property: Let $g \in G^{<0}$, $g \notin \dom \phi$.
If $g < g_0$, let $q < \min\{v_G(g), q_0\}$.
If $g > g_n$, we have $q_n < v_G(g_n) \leq v_G(g)$; so let $q_n < q < v_G(g)$.
If $g_i < g < g_{i+1}$, we have $q_i < v_G(g_i) \leq v_G(g)$; so let $q_i < q < \min\{v_G(g), q_{i+1}\}$.
In all cases, set $\phi(g) = q$.

Φ has the forth property: Let $q \notin \range \phi$, $q \in \mathbb{Q}$.
If $q < q_0$, let $q' \in \mathbb{Q}$ such that $q < q' < v_G(g_0)$.
If $q > q_n$, let $q' > \max\{v_G(g_n), q\}$.
If $q_i < q < q_{i+1}$ then $v_G(g_i) \leq v_G(g_{i+1})$ and $q < v_G(g_{i+1})$. Assume first that $v_G(g_i) < v_G(g_{i+1})$, then choose $q' \in \mathbb{Q}$ s.t. $\max\{v_G(g_i), q\} < q' < v_G(g_{i+1})$.
Now let $g \in G^{<0}$ such that $v_G(g) = q'$, in the above three cases. Finally, if $v_G(g_i) = v_G(g_{i+1})$, choose $g \in G^{<0}$ such that $g_i < g < g_{i+1}$ (here, we use the fact that G is dense).
In all cases, set $\phi(g) = q$. □

Remark 2.17 In a straightforward manner, the above procedure may be modified such that it produces a map

$$\tilde{\varphi} : G^{<0} \longrightarrow \mathbb{Q} \quad \text{satisfying} \quad \forall g : \tilde{\varphi}(g) > v_G(g) .$$

On the other hand, we can also achieve that "<" holds at some part of the group, while ">" holds at some other part. Indeed, we may partition $\mathbb{Q}$ into countably many disjoint open intervals I_j with $j \in J \subset \mathbb{N}$, and construct $\tilde{\varphi}$ such that for all

6. Countable exponential fields with growth properties   41

$g \in G$ with $v_G(g) \in I_j$, $\tilde{\varphi}(g) > v_G(g)$ whenever j is odd and $\tilde{\varphi}(g) < v_G(g)$ whenever j is even.

Corollary 2.18 *Assume that K admits an exponential and that $G = v(K)$ is countable. Then K admits a (GA)-v-left exponential.*

Proof If $G = 0$ then $\mathbf{A} = 0 = \mathbf{B}$ and there is nothing to prove. If $G \neq 0$ then it is of the form $G \simeq \coprod_{\mathbb{Q}} A$ as we have recalled at the beginning of this section. Hence, G has the lifting property (cf. the remark following Theorem 2.7). By the preceding proposition, G is a strong exponential group. The conclusion now follows from Theorem 2.7. $\square$

If K also is supposed to be countable, we may replace the condition that it admits an exponential by the condition that G is an exponential group in $(\overline{K}, +, 0, <)$. The following corollary gives a criterion for countable fields to admit an exponential satisfying the growth axioms.

Corollary 2.19 *Assume that K is a countable non-archimedean ordered field, root closed for positive elements. Assume that $\overline{K}$ is (isomorphic to) an exponentially closed subfield of $(\mathbb{R}, \exp)$. Assume further that G is an exponential group in $(\overline{K}, +, 0, <)$. Then K admits a (GA)-exponential.*

Proof Since $\overline{K}$ is exponentially closed, $\exp$ is defined on it. By Theorem 1.44, K admits an exponential f lifting $\exp$. Since K is countable, so is G. By Proposition 2.16, G admits a strong group exponential. Fix any decomposition (14) of $(K, +, 0, <)$ and let $f_L \amalg f_M \amalg f_R$ be the induced decomposition of f. By Theorem 2.7, K admits a (GA)-v-left exponential f'_L with the same domain and range as f_L. In particular the exponential $f' = f'_L \amalg f_M \amalg f_R$ satisfies that $f'(a) > a^n$ for all $n \geq 1$ (for $a > 0$ and $v(a) < 0$). Moreover, since $\overline{f'} = \overline{f} = \exp$ is a (GAT)-exponential on $\overline{K}$, we know by Theorem 2.9 that f' further satisfies: If $v(a) = 0$ and $a \geq n^2$ for some $n \geq 1$, then $f'(a) > a^n$. Thus by Remark 2.2, f' is a (GA)-exponential as required. $\square$

Thus for countable fields, admitting an exponential is "almost" equivalent to admitting a (GA)-exponential:

Corollary 2.20 *Assume K to be a countable non-archimedean exponential field, such that $\overline{K}$ is (isomorphic to) an exponentially closed subfield of $(\mathbb{R}, \exp)$. Then K admits a (GA)-exponential.*

It now remains to consider v-right exponentials in the countable case. By Corollary 1.39, we know that there exists an isomorphism $f : I_v \to 1 + I_v$ if K is countable and root closed for positive elements. But we want to realize additional conditions for that isomorphism.

Lemma 2.21 *Suppose that K is root closed for positive elements. Let $\mathcal{B}$ be a subset of I_v, $n \geq 1$ and*

$$\mathcal{B}_n = \{E_n(a) \mid a \in \mathcal{B}\} \subset 1 + I_v .$$

Then $\mathcal{B}$ is valuation independent in I_v if and only if $\mathcal{B}_n$ is valuation independent in $1 + I_v$, and $\mathcal{B}$ is maximal with this property if and only if $\mathcal{B}_n$ is. If $\mathcal{B}$ is valuation independent, then the map

$$\tilde{f}_n : \mathcal{B} \ni a \mapsto E_n(a) \in \mathcal{B}_n$$

extends additively to an order preserving isomorphism f_n from $\langle \mathcal{B} \rangle$ onto $\langle \mathcal{B}_n \rangle$ satisfying $Q_n(a, f_n(a))$ for all $a \in \langle \mathcal{B} \rangle$.

Proof Taking over the notation from Section 2 of Chapter 1, for "the smallest convex subgroup containing x" resp. "the biggest convex subgroup not containing x" we will write C_x resp. D_x for elements x in the additive group of K, and $\mathbf{C}_x$ resp. $\mathbf{D}_x$ for x in the multiplicative group of positive elements of K.

First note that for all $n \geq 1$ and all $a \in \mathcal{B}$,

$$v^{\cdot}(1+a) = v^{\cdot}(E_n(a)) \quad (\text{hence, } \mathbf{D}_{1+a} = \mathbf{D}_{E_n(a)})$$

and moreover,

$$(1+a) \cdot \mathbf{D}_{1+a} = E_n(a) \cdot \mathbf{D}_{1+a} \quad \left(\text{i.e., } v^{\cdot}\left(\frac{1+a}{E_n(a)}\right) > v^{\cdot}(1+a)\right).$$

In fact, by Theorem 1.16,

$$v^{\cdot}(1+a) = v(a) = v\left(\sum_{i=1}^{n} \frac{a^i}{i!}\right) = v^{\cdot}\left(1 + \sum_{i=1}^{n} \frac{a^i}{i!}\right) = v^{\cdot}(E_n(a))$$

and

$$\begin{aligned}
v^{\cdot}\left(\frac{1+a}{E_n(a)}\right) &= v\left(\frac{1+a}{E_n(a)} - 1\right) = v\left(\frac{1+a - E_n(a)}{E_n(a)}\right) \\
&= v\left(\sum_{i=2}^{n} \frac{a^i}{i!}\right) - v(E_n(a)) = 2v(a) \\
&> v(a) = v^{\cdot}(1+a).
\end{aligned}$$

So by Proposition 0.11 we see that for all $n \geq 1$, $\mathcal{B}_n$ is valuation independent in $1 + I_v$ if and only if $\mathcal{B}_1$ is. Similarly, by Corollary 0.12 we see that $\mathcal{B}_n$ is maximal with this property if and only if $\mathcal{B}_1$ is. So in order to prove the first assertion of our present corollary, we may assume w.l.o.g. that $n = 1$.

Now note that for given $a_1, \ldots, a_m \in \mathcal{B}$ we have: $v(a_1) = \ldots = v(a_m)$ if and only if $v^{\cdot}(1 + a_1) = \ldots = v^{\cdot}(1 + a_m)$. Moreover, in that case we have: $a_1 + D_{a_1}, \ldots, a_m + D_{a_m}$ are $\mathbb{Q}$–independent in C_{a_1}/D_{a_1} if and only if $(1 + a_1) \cdot \mathbf{D}_{1+a_1}, \ldots, (1+a_m) \cdot \mathbf{D}_{1+a_m}$ are $\mathbb{Q}$–independent in $\mathbf{C}_{1+a_1}/\mathbf{D}_{1+a_1}$. Indeed, this last statement is true because the map

$$\begin{aligned}
\phi_{1+a_1} : \mathbf{C}_{1+a_1}/\mathbf{D}_{1+a_1} &\to C_{a_1}/D_{a_1} \\
c \cdot \mathbf{D}_{1+a_1} &\mapsto (c-1) + D_{a_1}
\end{aligned}$$

is an order preserving isomorphism (cf. Lemma 1.14). Hence, Proposition 0.11 shows that $\mathcal{B}$ is valuation independent if and only if $\mathcal{B}_1$ is, and Corollary 0.12 shows that $\mathcal{B}$ is maximal with this property if and only if $\mathcal{B}_1$ is.

Finally, assume that $\mathcal{B}$ is valuation independent. Since $v(a) = v^{\cdot}(E_n(a))$, we know by Lemma 0.16 that f_n extends linearly to a valuation preserving isomorphism $\tilde{f}_n : \langle \mathcal{B} \rangle \to \langle \mathcal{B}_n \rangle$. By Lemma 2.13, it follows that $Q_n(a, \tilde{f}_n(a))$ holds for all $a \in \langle \mathcal{B} \rangle$. Consequently, if $a \in \langle \mathcal{B} \rangle$, $a > 0$, then by virtue of part i) of Lemma 2.10 we have $\tilde{f}_n(a) > E_{n-1}(a) \geq 1$, which shows that f_n preserves the order. □

Call a v-right exponential a $(\mathbf{T}_n)$-v-**right exponential** if it satisfies the axioms $(\mathbf{T}_m)$ for all $1 \leq m \leq n$ on its domain.

Theorem 2.22 *Suppose that K is a countable field, root closed for positive elements. Then for $n = 1, 2$, there exists a right exponential $f_n : I_v \to 1 + I_v$ and a valuation basis $\{a_j \mid j \in \mathbb{N}\}$ (depending on n) of I_v such that:*
(1) $f_n(a_j) = E_n(a_j)$ for all $j \in \mathbb{N}$,
(2) $\{f_n(a_j) \mid j \in \mathbb{N}\}$ is a valuation basis of $1 + I_v$,
(3) $Q_n(x, f_n(x))$ holds for all $x \in I_v$, $x \neq 0$ and thus, f_n is a (T_{n-1})-v-right exponential.

If in addition, K is henselian for its natural valuation v (which in particular is the case if K is real closed), then there is such a v-right exponential f_n for all n.

Proof First note that the conclusion about f_n in 3) follows by Lemma 2.10. We will construct the required isomorphism by a back and forth procedure. Since K is assumed to be countable, I_v admits a countable basis $\{a'_j \mid j \in \mathbb{N}\}$ and $1 + I_v$ admits a countable basis $\{b'_j \mid j \in \mathbb{N}\}$. Given $m \geq 0$, assume that we have already constructed an isomorphism $f_{n,m}$ between an m-dimensional subvector space U_m of I_v and a subvector space V_m of $1 + I_v$ and a v-valuation basis $\{a_1, \ldots, a_m\}$ of U_m such that
$(1)_m$ $f_{n,m}(a_j) = E_n(a_j)$ for $1 \leq j \leq m$,
$(2)_m$ $\{f_{n,m}(a_j) \mid 1 \leq j \leq m\}$ is a $v\dot{}$-valuation basis of V_m,
$(3)_m$ $Q_n(x, f_{n,m}(x))$ holds for all $x \in U_m$, $x \neq 0$.

By Lemma 2.21, conditions $(2)_m$ and $(3)_m$ both follow from $(1)_m$. By convention, for $m = 0$ we set $U_0 = \{0\}$ and $V_0 = \{1\}$. If m is even, then let a be the basis element a'_j of smallest index j such that $a'_j \notin U_m$. If m is odd, then let b be the basis element b'_j of smallest index j such that $b'_j \notin V_m$. We want to construct an extension $f_{n,m+1}$ of $f_{n,m}$ to $U_m + \mathbb{Q}a$ resp. of $f_{n,m}^{-1}$ to $V_m \cdot b^{\mathbb{Q}}$ still satisfying condition $(1)_{m+1}$ on the new domains. By Lemma 0.13, we may extend the v-valuation basis $\{a_1, \ldots, a_m\}$ of U_m to a v-valuation basis $\{a_1, \ldots, a_m, a_{m+1}\}$ of $U_m + \mathbb{Q}a$ (respectively, the $v\dot{}$-valuation basis $\{f_{n,m}(a_1), \ldots, f_{n,m}(a_m)\}$ of V_m to a $v\dot{}$-valuation basis $\{f_{n,m}(a_1), \ldots, f_{n,m}(a_m), b_{m+1}\}$ of $V_m \cdot b^{\mathbb{Q}}$).

If m is even, we take $f_{n,m+1}(a_{m+1})$ to be equal to $E_n(a_{m+1})$, which we will call d. If m is odd, we are looking for an element $c \in I_v$ such that $E_n(c) = b_{m+1}$. For $n = 1$, this is just $c = b_{m+1} - 1$. For $n = 2$, our task requires to solve an equation
$$X^2 + 2X + 2(1 - b_{m+1}) = 0$$
which is always solvable in the root-closed field K since
$$1 - 2(1 - b_{m+1}) = 2b_{m+1} - 1 > 0$$
in view of $b_{m+1} \in 1 + I_v$. If $n > 2$, the equation
$$\sum_{i=0}^{n} \frac{1}{i!} X^i - b_{m+1} = 0$$
is still solvable if the natural valuation on K is henselian. Indeed, $v(1 - b_{m+1}) > 0$ and hence, the above equation will then admit a root c whose residue is 0, that is, $c \in I_v$.

Now note that by Lemma 2.21, $\{a_1, \ldots, a_m, c\}$ is again v-valuation independent (resp. $\{f_{n,m}(a_1), \ldots, f_{n,m}(a_m), d\}$ is again $v\dot{}$-valuation independent). Hence, we set $a_{m+1} := c$ resp. $b_{m+1} := d$ and obtain in both cases that $f_{n,m}(a_{m+1}) = b_{m+1}$. So indeed, we are able to extend $f_{n,m}$ to an isomorphism $f_{n,m+1}$ such that $a \in \text{dom } f_{n,m+1}$ resp. $b \in \text{im } f_{n,m+1}$.

We set $f_n = \bigcup_{m\in\mathbb{N}} f_{n,m}$. Since by our back and forth construction, every a'_j is contained in some U_m and every b'_j is contained in some V_m, we find that f_n is an isomorphism from I_v onto $1 + I_v$. Since for every $m \in \mathbb{N}$, the isomorphism $f_{n,m}$ has the properties $(1)_{m+1}$, $(2)_{m+1}$ and $(3)_{m+1}$, the induced isomorphism $f_{n,m}$ has the properties (1), (2) and (3). Finally, since every finite subset $\{a_1, \ldots, a_m\}$ of $\{a_j \mid j \in \mathbb{N}\}$ is a valuation basis of U_m, the set $\{a_j \mid j \in \mathbb{N}\}$ itself is a valuation basis of $I_v = \bigcup_{m\in\mathbb{N}} U_m$. □

Finally, let us put left, middle and right together.

Theorem 2.23 *Let K be a countable non-archimedean ordered field, root closed for positive elements. Assume that its value group is isomorphic to $\coprod_\mathbb{Q}(\overline{K}, +, 0, <)$ and that its residue field $\overline{K}$ admits a (GAT)-exponential e. Then K admits a (GAT_1)-exponential f lifting e.*

If in addition, K is henselian for its natural valuation, then for every fixed $n \geq 1$, K admits a (GAT_n)-exponential f lifting e.

Proof We may assume the decompositions (14) and (15) since K is root closed for positive elements. If $v(K) \simeq \coprod_\mathbb{Q} \overline{K}$, then by Corollary 2.18, K admits a (GA)-v-exponential f_L. Further, the exponential e of $\overline{K}$ induces a (GAT)-v-middle exponential. From Theorem 2.22 we infer the existence of a (T_1)-v-right exponential f_R. Under the additional assumption that K be henselian for its natural valuation, we may replace (T_1) by (T_n) for an arbitrary fixed n. By virtue of Lemma 2.14, the resulting exponential $f = f_L \amalg f_M \amalg f_R$ is a (GAT_1) respect. a (GAT_n)-exponential on K, lifting $\overline{f} = e$. □

Let us mention that the last theorem is the best that can be expected, in the following sense. If we would try to obtain a (GAT)-exponential, we would have to construct a (T)-v-right exponential, that is, an exponential f_R satisfying $\forall x \in I_v : Q_n(x, f_R(x))$ simultaneously for all $n \in \mathbb{N}$. With our approach, the existence would only be guaranteed if we would ensure some convergence, but this usually contradicts the countability condition. Certainly, countable fields with (GAT)-exponentials can be constructed in a different way, but not in the sense of giving a criterion for a countable field to admit a (GAT)-exponential.

Note that a (GAT_1)-exponential is continuous, differentiable and equal to its own derivative (cf. Theorem 14 of [DA–WO]). We have thus shown the following:

Corollary 2.24 *Let K be a countable real closed field with exponentially closed (in $(\mathbb{R}, \exp)$) residue field $\overline{K}$. Then $\exp$ can be lifted to a continuous, differentiable exponential f on K satisfying the differential equation $f' = f$, provided that $G \simeq \coprod_\mathbb{Q}(\overline{K}, +, 0, <)$.*

7 Natural contractions arising from logarithms

The disadvantage of the group exponential $\tilde{h}$ that we have used so far is that it is not a map from G to G. But if we compose it with the natural valuation $v_G : G \to v_G(G)$, we obtain a very interesting map $\chi : G^{<0} \to G^{<0}$, which we now want to analyze.

Let G be an ordered abelian group and χ a map from $G^{<0}$ into $G^{<0}$. Then χ will be called a **contraction** if it satisfies the following axioms:

(C1) χ is surjective,

(C2) χ preserves $\leq$,

7. Natural contractions arising from logarithms

(C3) if g is archimedean equivalent to g', then $\chi(g) = \chi(g')$.

Note that for negative group elements: g is archimedean equivalent to g' if and only if there is $n \in \mathbb{N}$ such that $ng < g'$ and $ng' < g$. Similarly, g is infinitely smaller than g' if and only if for all $n \in \mathbb{N}$ we have $ng > g'$. In the language of ordered abelian groups, axiom (C3) is not an elementary sentence, but in the presence of (C2) it is equivalent to a (recursive) axiom scheme:

(C3') $\forall x, y : \ x \leq y < 0 \wedge ny \leq x \ \rightarrow \ \chi(x) = \chi(y) \qquad (n \in \mathbb{N})$.

We will call χ a **natural contraction** if $\chi(x) = \chi(y)$ implies that x and y are archimedean equivalent. This notion is not elementary: by general model theory, it can be shown that in every $\aleph_0$–saturated model, the contraction will contract elements that are not archimedean equivalent.

Natural contractions are in a one-to-one correspondence with group exponentials as described in the following two lemmas. The proofs of the lemmas are left to the reader.

Lemma 2.25 *Let $\tilde{h} : v_G(G) \to G^{<0}$ be a group exponential on G. Define $\chi : G^{<0} \to G^{<0}$ by*

$$\chi = \tilde{h} \circ v_G.$$

Then χ is a natural contraction on G.

We call $\chi = \tilde{h} \circ v_G$ the **natural contraction induced by $\tilde{h}$**.

Lemma 2.26 *Let χ be a contraction on G. Define $\tilde{h} : v_G(G) \to G^{<0}$ by*

$$\tilde{h}(v_G(g)) = \chi(g)$$

(for $g \in G^{<0}$). Then $\tilde{h}$ is well-defined, surjective and order preserving. Moreover, $\tilde{h}$ is an isomorphism (that is, a group exponential) if and only if χ is a natural contraction.

We call $\tilde{h}$ thus obtained the **group exponential induced by χ**.

Our main example comes naturally from exponential fields:

Example 2.27 Let f be a v-compatible exponential on K, set $\ell = f^{-1}$. Let

$$\chi_f = \tilde{h}_f \circ v_G$$

be the natural contraction induced by the group exponential $\tilde{h}_f$ (cf. Lemma 2.25). We call χ_f the **natural contraction induced by f**.

We wish to compute $\chi_f(g)$ for $g \in G^{<0}$ in terms of ℓ. To do this, fix any decomposition (14) of $(K, +, 0, <)$. Let $f_L : \mathbf{A} \to \mathbf{B}$ be the induced v-left exponential, and $h_f : G \to \mathbf{A}$ the induced isomorphism. Choose $a > 0$ such that $v(a) = g$, and let $b \in \mathbf{B}$ be the uniquely determined element for which $v(b) = g$. Then

$$\begin{aligned} \chi_f(g) &= = \tilde{h}_f(v_G(g)) = v(h_f(g)) = v((\ell \circ (-v)^{-1})(v(b))) \\ &= v(\ell(b^{-1})) = v(-\ell(b)) = v(\ell(b)) \,. \end{aligned}$$

Now $a = bu$ with u a unit, so $\ell(a) = \ell(b) + \ell(u)$. Thus $v(\ell(a)) = v(\ell(b) + \ell(u))$. Now $v(\ell(b)) < 0$ and $v(\ell(u)) \geq 0$ by v-compatibility. It follows by the ultrametric inequality that $v(\ell(a)) = v(\ell(b))$.

We conclude:

$$\chi_f(v(a)) = v(\ell(a)) \quad \text{for all } a > 0 \text{ with } v(a) < 0.$$

Note that χ_f depends only on f and *not* on the chosen decomposition. Note also that natural contractions are already induced by v-left exponentials. The only disadvantage there is that the defining formula for χ_f is not as straightforward: we have
$$\chi_f(v(a)) = v(\ell(b))$$
for all $a > 0$ with $v(a) < 0$, where $b \in \mathbf{B}$ is the uniquely determined element for which $v(b) = v(a)$.

We would now like to encode the (GA)-axiom scheme in the behaviour of the contraction, and to study the relation to strong groups exponentials. Clearly, if $\tilde{h}$ is a strong group exponential on G then the induced contraction satisfies
$$\forall g \in G^{<0} : g < \chi(g).$$
That is, χ maps towards the center of the ordered group (which is the element 0). This gives rise to the following definitions: a contraction χ is **centripetal**, if it satisfies

(CP) $\forall g \in G^{<0} : g < \chi(g)$,

and it will be called **centrifugal**, if it satisfies

(CF) $\forall g \in G^{<0} : g > \chi(g)$.

The proof of the following observation is straightforward:

Lemma 2.28 *Let $(G, \tilde{h})$ be an exponential group and χ the contraction induced by $\tilde{h}$. Then:*

a) χ is centripetal if and only if $(G, \tilde{h})$ is a strong exponential group.

b) χ is centrifugal if and only if $\tilde{h}$ satisfies
$$\forall g \in G^{<0} : \tilde{h}(v_G(g)) < g . \tag{25}$$

Corollary 2.29 *Let f be a v-left exponential on K. Then f is a (GA)-exponential if and only if the induced contraction χ_f is centripetal.*

In view of this lemma, Proposition 2.16 has shown that the group $\coprod_\mathbb{Q} A$ can be endowed with a group exponential inducing a centripetal contraction. By Remark 2.17, it also admits a group exponential inducing a centrifugal contraction, as well as a group exponential inducing a contraction which is neither centrifugal nor centripetal. This shows that axioms (CP) and (CF) are independent of the other contraction axioms.

Property (25) reflects a quite strange behaviour of an exponential. Indeed, if χ_f induced by an exponential (or a left exponential) f on K satisfies (25), then
$$\forall x \in K : \ v(x) < 0 \to v_G(v(f(x))) > v_G(v(x)) .$$

This in turn means that for infinitely big $a \in K$ we have that $f(a) \ll a$, that is, $f(a)$ is smaller than any root of a. In the presence of a predicate for the natural valuation, it is possible to axiomatize such strange exponential fields where the exponential induces the usual exponential on the "finite" part of the field (the convex hull of $\mathbb{Q}$, which is the valuation ring of the natural valuation), but "reverses" its behaviour on some "infinite" part of the field (some infinitely big elements). However, since the natural valuation cannot be axiomatized elementarily, it is not possible to axiomatize elementarily a class of exponential fields where this reversed behaviour is shown at *all* infinitely big elements.

Abelian groups with contractions have a fascinating model theory, which has been worked out in detail by F.- V. Kuhlmann. These results are presented in the Appendix of this monograph.

CHAPTER 3

The exponential rank

Based on the work of Hahn, Baer, Ostrowski, Krull, Kaplansky and the Artin-Schreier theory, and stimulated by the paper [LAN] in 1953, the theory of real places and convex valuations has witnessed a remarkable development. It has become a basic tool in the theory of ordered fields and real algebraic geometry. Surveys on this development can be found in [LAM2] and [PC]. In this chapter, we take a further step by adding an exponential function to the ordered field. Beforehand, we sketch the basic facts about convex valuations.

1 Convex valuations

Throughout this chapter, K will be a non-archimedean ordered field, and v will denote its non-trivial natural valuation. We set: $G = v(K)$ and $\Gamma = v_G(v(K)) = v_G(G^{<0})$.

Let w be a valuation of K, with valuation ring R_w, valuation ideal I_w, value group $w(K)$ and residue field Kw. Then w is called **compatible with the order** if and only if it satisfies, for all $a, b \in K$:

(CO) $\qquad a \geq b > 0 \implies w(a) \leq w(b)$.

For every valuation w of a field K and every $a \in K$, we have that $w(a) = w(-a)$ and that $w(0) \geq w(a)$. Hence it follows directly from our definition that every order compatible valuation w satisfies:

$$a \leq b \leq 0 \vee a \geq b \geq 0 \implies w(a) \leq w(b) .$$

For the following characterizations of compatible valuations, cf. [LAM1], Proposition 5.1, or [LAM2], Theorem 2.3 and Proposition 2.9, or [PR1], Lemma 7.2:

Lemma 3.1 *The following assertions are equivalent:*
1) *w is a valuation compatible with the order of $(K, <)$,*
2) *R_w is a convex subset of $(K, <)$,*
3) *I_w (or equivalently, the set $1 + I_w$ of 1-units) is a convex subset of $(K, <)$,*
4) *the positive cone of $(K, <)$ contains $1 + I_w$,*
5) *$I_w < 1$,*
6) *the image of the positive cone of $(K, <)$ under the residue map*
   $K \ni a \mapsto aw \in Kw$ is a positive cone in Kw.

Thus by Lemma 3.1 above, w is compatible with the order if and only the order of K canonically induces an order on the residue field Kw. This in turn holds if and only if R_w and I_w are convex subsets of $(K, <)$. Therefore, a valuation compatible with the order is also said to be a **convex valuation**. For every convex valuation w, the set

$$\mathcal{U}_w^{>0} := \{a \in K \mid w(a) = 0 \wedge a > 0\}$$

of **positive units** of R_w is a convex subgroup of the ordered multiplicative group $(K^{>0}, \cdot, 1, <)$ of positive elements of K.

Let w and w' be valuations on K. We say that w' is **finer** than w, or that w is **coarser** than w' if $R_{w'} \subsetneq R_w$. This is equivalent to $I_w \subsetneq I_{w'}$. If w' is convex, it follows through assertion 4) or 5) that w also is convex. If w' is finer than w, then there is a non-trivial valuation w'/w on Kw. If w' is convex, then Kw has an induced order with respect to which w'/w is convex.

Since the residue field of the natural valuation v with the induced order is archimedean ordered, it follows that there is no convex valuation finer than v. Thus the natural valuation v of K is the finest convex valuation. It is characterized by the fact that its residue field is an archimedean ordered field.

By general valuation theory, the set $\mathcal{R}$ of all valuation rings R_w of convex valuations $w \neq v$ is totally ordered by inclusion. Its order type is called the **rank** of $(K, +, \cdot, 0, 1, <)$. For convenience, we will identify it with $\mathcal{R}$.

Example 3.2 The rank of an archimedean ordered field is empty since its natural valuation is trivial. The rank of the rational function field $K = \mathbb{R}(t)$ with any order is a singleton: $\mathcal{R} = \{K\}$.

To every convex valuation ring R_w, we associate a convex subgroup G_w of $v(K)$:
$$G_w := \{v(a) \mid a \in K \wedge w(a) = 0\} = v(\mathcal{U}_w^{>0}).$$

The value group $w(K)$ is canonically isomorphic to $v(K)/G_w$. We call G_w the **convex subgroup associated to** w. For example, $G_v = \{0\}$. Conversely, given a convex subgroup G_w of $v(K)$, we define $w : K \to v(K)/G_w$ by $w(a) = v(a) + G_w$. Then w is a convex valuation with $v(\mathcal{U}_w^{>0}) = G_w$ (and v is finer than w if and only if $G_w \neq \{0\}$). We call w the **convex valuation associated to** G_w. We summarize (cf. [PC] for details):

Lemma 3.3 *Let v, w be arbitrary valuations on some field K. Suppose that v is finer than w. Then for all $a, b \in K$,*
$$v(a) \leq v(b) \Rightarrow w(a) \leq w(b). \tag{26}$$
In particular, $w(a) > 0 \Rightarrow v(a) > 0$. Let $G_w := \{v(z) \mid z \in K \wedge w(z) = 0\} \subset v(K)$ be the convex subgroup associated to w. We have that $v(z) \in G_w \Leftrightarrow z \in \mathcal{U}_w$. There is a canonical isomorphism $w(K) \simeq v(K)/G_w$. Conversely, every convex subgroup C of $v(K)$ is of the form G_w where w is the convex valuation associated to C. The valuation v of K induces a valuation v/w on Kw. There are canonical isomorphisms $(v/w)(Kw) \simeq G_w$ and $(Kw)(v/w) \simeq Kv$. If Kw is embedded in R_w such that the restriction of the residue map is the identity on Kw, then $v/w = v|Kw$ (up to equivalence). Writing v instead of $v|Kw$, we then have that $v(Kw) = G_w$ and $(Kw)v = Kv$.

Recall from Chapter 0 that the set of all convex subgroups $G_w \neq \{0\}$ of the value group G is totally ordered by inclusion. Its order type is called the **rank** of G.

Note that the rank of an ordered field (respectively of an ordered group) is an invariant.

Lemma 3.4 *The correspondence $R_w \mapsto G_w$ is one-to-one and order preserving, thus $\mathcal{R}$ is (isomorphic to) the rank of G.*

Thus we want to analyze the rank of G further and relate it to another invariant that we have encountered already, namely to the value set Γ of G.

1. Convex valuations

The proof of the following lemmas is straightforward.

Lemma 3.5 *Let G be an ordered Abelian group.*
(i) If $G_w \neq \{0\}$ is a convex subgroup, then
$$\Gamma_w := v_G(G_w)$$
is a non-empty final segment of Γ.
(ii) Conversely, if Γ_w is a non-empty final segment of Γ, then
$$G_w = \{g \mid g \in G, v_G(g) \in \Gamma_w\} \cup \{0\}$$
is a convex subgroup, with $\Gamma_w = v_G(G_w)$.

Let us denote by Γ^{fs} the set of non-empty final segments of Γ, totally ordered by inclusion. Clearly the correspondence $G_w \mapsto \Gamma_w$ is one-to-one and order preserving, thus the rank of G is (isomorphic to) Γ^{fs}.

Lemma 3.6 *The map from Γ to Γ^{fs} defined by*
$$\gamma \mapsto \{\gamma' \mid \gamma' \in \Gamma, \gamma' \geq \gamma\}$$
is an order reversing embedding. Its image consists of those final segments which have a smallest element.

Thus if Γ^* denotes the set Γ with its reversed ordering, then the map given in Lemma 3.6 is an order *preserving* embedding of Γ^* into Γ^{fs}. A final segment which has a smallest element is a **principal final segment**. Thus Γ^* is (isomorphic to) the totally ordered set of principal final segments.

Recall from Chapter 0 that a convex subgroup G_w of G is called principal if there is some $g \in G$ such that G_w is the minimal convex subgroup containing g (it exists since the intersection of all convex subgroups containing g is a convex subgroup). In this case we say that G is **principal generated by** g.

Lemma 3.7 *Let $G_w \neq \{0\}$ be a convex subgroup. Then the following are equivalent:*
(i) G_w is principal convex (generated by g).
(ii) The archimedean class $[g]$ is the smallest amongst all classes of G_w (if and only if $[g] \cap G_w^{<0}$ is an initial segment in G_w, if and only if $[g] \cap G_w^{>0}$ is a final segment in G_w).
(iii) $v_G(G_w) = \Gamma_w$ is a principal final segment.

By the **principal rank** of G we mean the subset of the rank of G consisting of all principal $G_w \neq \{0\}$.

Lemma 3.8 *The map $G_w \mapsto \min v_G(G_w)$ is an order reversing bijection between the principal rank and Γ. Thus the principal rank is (isomorphic to) Γ^*.*

Note that under the bijective correspondence $G_w \mapsto \Gamma_w$ given in Lemma 3.5, the principal rank is precisely the preimage of the set of principal final segments. Note also that the correspondence $R_w \mapsto \Gamma_w$ is bijective between the rank of K and Γ^{fs}.

A convex subring $R_w \neq R_v$ is a **principal convex subring generated by** a if $a \in \mathbf{P}_K$ such that R_w is the smallest convex subring containing a (it exists since the intersection of all convex subrings containing a is a convex subring). By the **principal rank** of K we mean the subset $\mathcal{R}^{\mathrm{pr}}$ of $\mathcal{R}$ consisting of all principal $R_w \in \mathcal{R}$.

Recall that $\dot\sim$ denotes the multiplicative equivalence relation on $\mathbf{P}_K$. Denote the class of a by $[a]^{\cdot}$, and call it the **multiplicative class** of a (see figure).

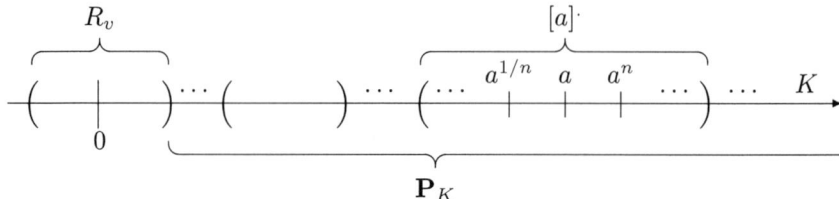

By Lemma 1.10 we get:

Theorem 3.9 *Let $R_w \neq R_v$ be a convex subring. Then the following are equivalent:*
(i) R_w is principal convex (generated by a).
(ii) The class $[a]^{\cdot}$ is the largest amongst all classes of R_w (which is equivalent to $[a]^{\cdot}$ is a final segment in R_w).
(iii) G_w is principal (with smallest archimedean class $[v(a)]$).
(iv) $v_G(G_w) = \Gamma_w$ is a principal final segment (with smallest element $v_G(v(a))$.
Thus the principal rank of K is (isomorphic to) Γ^*.

A corollary to the above is the following:

Theorem 3.10 *Let $w \neq v$ be a convex valuation of K. Then R_w lies in the principal rank if and only if for some b in the residue field Kw, the sequence $(b^n)_{n \in \mathbb{N}}$ is cofinal in Kw. The principal rank is a singleton (namely, $\mathcal{R}^{\mathrm{pr}} = \{K\}$) if and only if $(a^n)_{n \in \mathbb{N}}$ is cofinal in K for every positive infinite element $a \in K$.*

2 The exponential analogue of the rank

In the next sections, we will develop an exponential analogue of what has been reviewed above. That is, we develop a theory of "exponential places" and exponential compatible valuations, in analogy to the theory of real places and order compatible valuations. In the exponential case, the exponential function will play the role that multiplication plays in the real case.

Let f be an exponential on K, and w be a convex valuation on K. Then we will say that w and f are compatible or that f is a w-**compatible exponential** if the following holds:

(CE) $\qquad f(R_w) = \mathcal{U}_w^{>0}$ and $f(I_w) = 1 + I_w$.

If $\ell = f^{-1}$ is the logarithm, then it is compatible with w if and only if $\ell(\mathcal{U}_w^{>0}) = R_w$ and $\ell(1 + I_w) = I_w$.

Since $Kw = R_w/I_w$ and $(Kw^{>0}, \cdot, 1, <) = (\mathcal{U}_w^{>0}, \cdot, 1, <)/1 + I_w$, this means that f induces canonically an exponential

$$fw : (Kw, +, 0, <) \to (Kw^{>0}, \cdot, 1, <)$$

on the residue field Kw. (This is the analogue to the characteristic property of convex valuations to canonically induce an order on the residue field.) Recall that if an ordered field K admits any exponential, then it admits an exponential compatible with the natural valuation (cf. Lemma 1.18). Therefore, **we will assume throughout this chapter that every appearing exponential is v-compatible.** The valuation rings R_w of convex valuations $w \neq v$ satisfying the *first condition* of (CE) form a subset $\mathcal{R}_f$ of $\mathcal{R}$. Its order type will be called the **exponential rank** of the exponential field $(K, <, f)$; again, we identify it with $\mathcal{R}_f$. We shall see at the end of the next section that $\mathcal{R}_f$ coincides with the set of convex valuations compatible with f (if f is a (GA)-(T_1)-exponential).

We wish to characterize the corresponding convex subgroups G_w. However, these subgroups do not carry any information concerning the second condition of (CE). So it may well happen that $\mathcal{R}_f$ also contains valuation rings of valuations which are not compatible with f. A natural way to overcome this deficiency is to require that f satisfies the Taylor axiom:

(T$_1$) $\qquad v(f(a) - 1 - a) > v(a) \qquad$ for all $a \in I_v$.

(see Chapter 2 where this valuation theoretic formulation of the Taylor axioms was derived). Recall that if (T$_1$) holds, then we call f a **(T$_1$)-exponential**.

If f satisfies (T$_1$), then $f(I_w) = 1 + I_w$ holds for every convex valuation w (cf. Lemma 3.17 below). Then f is compatible with w if and only if $f(R_w) = \mathcal{U}_w^{>0}$ and consequently, $\mathcal{R}_f$ is precisely the set of all valuation rings R_w of valuations $w \neq v$ which are compatible with f.

We shall characterize the subgroups G_w of $v(K)$ for which $R_w \in \mathcal{R}_f$ by use of the contraction χ_f induced on $v(K)$ by the exponential f (more precisely, by its inverse, the logarithm ℓ, cf. Chapter 2). For the details, see the next section.

To avoid unpleasant case distinctions which would make the theory complicated without telling anything more about the interesting cases, we fix the "orientation" of these two maps. This is done by requiring that f satisfies (GA). Thus we will work with (GAT$_1$)-exponentials and -logarithms.

At this point, we have reached a crucial turn in this monograph. Indeed, in Chapter 4 we shall show that no power series field admits a surjective logarithm. On the other hand, we show that every power series field admits a non-surjective logarithm, and we will use this fact in Chapter 5 to construct models of real exponentiation. Thus we now have to learn to work with non-surjective logarithms.

3 (GA)- and (T$_1$)-prelogarithms

Let K be an ordered field. A **prelogarithm** on K is an embedding of the ordered Abelian group $(K^{>0}, \cdot, 1, <)$ into the ordered Abelian group $(K, +, 0, <)$. Thus a surjective prelogarithm is just a logarithm.

For a prelogarithm ℓ we will say that w **and** ℓ **are compatible** or that ℓ is a w-**compatible prelogarithm** if

$$\ell(\mathcal{U}_w^{>0}) = R_w \cap \mathrm{im}(\ell) \quad \text{and} \quad \ell(1 + I_w) = I_w \cap \mathrm{im}(\ell) \, . \tag{27}$$

As for exponentials, **we will assume throughout this chapter that all appearing prelogarithms are v-compatible.**

We let $\mathcal{R}_\ell$ denote the subset of $\mathcal{R}$ containing all R_w for which w satisfies the first condition of (27), and we call it the **exponential rank of** ℓ. If $\ell = f^{-1}$, then $\mathcal{R}_\ell = \mathcal{R}_f$.

Lemma 3.11 *Let ℓ be a prelogarithm on K. Then $\ell(\mathcal{U}_w^{>0}) = R_w \cap \mathrm{im}(\ell)$ if and only if*

$$\ell(\mathcal{U}_w^{>0}) \subset R_w \quad \text{and} \quad \ell(K^{>0} \setminus \mathcal{U}_w^{>0}) \subset K \setminus R_w \, . \tag{28}$$

Proof First assume that $\ell(\mathcal{U}_w^{>0}) = R_w \cap \mathrm{im}(\ell)$. Then clearly $\ell(\mathcal{U}_w^{>0}) \subset R_w$. We show that $\ell(K^{>0} \setminus \mathcal{U}_w^{>0}) \subset K \setminus R_w$. If not, there is $x \in K^{>0} \setminus \mathcal{U}_w^{>0}$ such that $\ell(x) \in R_w$. Thus $\ell(x) \in R_w \cap \mathrm{im}(\ell)$. Thus $\ell(x) \in \ell(\mathcal{U}_w^{>0})$, which is a contradiction since $x \notin \mathcal{U}_w^{>0}$.

Now assume that (28) holds. Then clearly $\ell(\mathcal{U}_w^{>0}) \subset R_w \cap \text{im}(\ell)$. If $y \in R_w \cap \text{im}(\ell)$, then $y = \ell(x)$ for some $x > 0$. Clearly $x \in \mathcal{U}_w^{>0}$, for if $x \in K^{>0} \setminus \mathcal{U}_w^{>0}$, then $\ell(x) \notin R_w$ which is a contradiction. $\square$

We now make a further reduction.

Lemma 3.12 *Let ℓ be a prelogarithm on K. Then $\ell(K^{>0} \setminus \mathcal{U}_w^{>0}) \subset K \setminus R_w$ if and only if*

$$\ell(K^{>0} \setminus R_w) \subset K^{>0} \setminus R_w . \tag{29}$$

Proof We first claim that $\ell(K^{>0} \setminus \mathcal{U}_w^{>0}) \subset K \setminus R_w$ if and only if

$$a \in K^{>0} \wedge w(a) < 0 \Rightarrow w(\ell(a)) < 0 . \tag{30}$$

Indeed let $a \in K^{>0} \setminus \mathcal{U}_w^{>0}$. If $a \notin R_w$ then $w(a) < 0$ thus $w(\ell(a)) < 0$ so $\ell(a) \in K \setminus R_w$. If $a \in R_w$ then necessarily $a \in I_w$. But $w(a) > 0 \Leftrightarrow w(a^{-1}) < 0$. Thus $0 > w(\ell(a^{-1})) = w(-\ell(a)) = w(\ell(a))$. Thus $\ell(a) \in K \setminus R_w$. The other direction of the claim is straightforward.

Now clearly (29) implies (30). Conversely (30) implies (29). Just note that $a \in K^{>0} \setminus R_w$ implies that $a > 1$ and thus, $\ell(a) > \ell(1) = 0$. This ends the proof. $\square$

We now consider the content of (GA) and (T_1) for prelogarithms. We recall briefly from Chapter 2 some facts that were proved for logarithms, but which hold for prelogarithms as well. As we have seen, (GA) is only relevant for $v(a) \leq 0$. Now we are interested in the case of $v(a) < 0$. In this case, "$a > n^2$" holds for all $n \in \mathbb{N}$ if a is positive. Restricted to $K \setminus R_v$, axiom scheme (GA) is thus equivalent to the assertion

$$\forall n \in \mathbb{N}: f(a) > a^n \qquad \text{for all } a \in K^{>0} \setminus R_v . \tag{31}$$

Applying the logarithm $\ell = f^{-1}$ on both sides, we find that this is equivalent to

$$\forall n \in \mathbb{N}: a > \ell(a^n) = n\ell(a) \qquad \text{for all } a \in K^{>0} \setminus R_v . \tag{32}$$

Via the natural valuation v, this in turn is equivalent to

$$v(a) < v(\ell(a)) \qquad \text{for all } a \in K^{>0} \setminus R_v . \tag{33}$$

A prelogarithm ℓ will be called a **(GA)-prelogarithm** if it satisfies (32).

Lemma 3.13 *Assume that w is a convex valuation, and ℓ a (GA)-prelogarithm on K. Then*

$$\ell(\mathcal{U}_w^{>0}) \subset R_w.$$

Proof Since $R_w \supset R_v$ we have that $v(a) < v(b)$ implies $w(a) \leq w(b)$. Hence, (33) implies:

$$w(a) \leq w(\ell(a)) \qquad \text{for all } a \in K^{>0} \setminus R_v . \tag{34}$$

This in turn implies that

$$\ell(\mathcal{U}_w^{>0} \setminus R_v) \subset R_w . \tag{35}$$

We note that

$$\mathcal{U}_w^{>0} = (\mathcal{U}_w^{>0} \cap I_v) \cup \mathcal{U}_v^{>0} \cup (\mathcal{U}_w^{>0} \setminus R_v) \tag{36}$$

with

$$(\mathcal{U}_w^{>0} \cap I_v) < \mathcal{U}_v^{>0} < (\mathcal{U}_w^{>0} \setminus R_v) \tag{37}$$

3. (GA)- and (T_1)-prelogarithms

and
$$\mathcal{U}_w^{>0} \cap -R_v = \{a^{-1} \mid a \in \mathcal{U}_w^{>0} \setminus R_v\}. \tag{38}$$

Since ℓ is compatible, we have that $\ell(\mathcal{U}_v^{>0}) \subset R_v \subset R_w$. Since ℓ also satisfies (35), then $\ell(\mathcal{U}_w^{>0} \setminus R_v) \subset -R_w = R_w$, and
$$\ell(\mathcal{U}_w^{>0}) = \ell(\mathcal{U}_w^{>0} \cap I_v) \cup \ell(\mathcal{U}_v^{>0}) \cup \ell(\mathcal{U}_w^{>0} \setminus R_v) \subset R_w, \tag{39}$$
as required. □

A corollary to the last three lemmas is

Lemma 3.14 *Assume that ℓ is a (GA)-prelogarithm. Then $\ell(\mathcal{U}_w^{>0}) = R_w \cap \mathrm{im}(\ell)$ (that is, $R_w \in \mathcal{R}_\ell$) if and only if*
$$\ell(K^{>0} \setminus R_w) \subset K^{>0} \setminus R_w.$$

Let us note the following important corollary to the proof of Lemma 3.13, for future use:

Corollary 3.15 *Let $w \neq v$ be a convex valuation on K, and let v_G denote the natural valuation on $v(K)$. Then*
$$v_G(v(\mathcal{U}_w^{>0})) = v_G(\{v(a) \mid a \in R_w^{>0} \setminus R_v^{>0}\}).$$

Proof Since
$$\mathcal{U}_w^{>0} \setminus R_v = R_w^{>0} \setminus R_v^{>0},$$
(36), (37) and (38) imply that
$$\mathcal{U}_w^{>0} = \{a^{-1} \mid a \in (R_w^{>0} \setminus R_v^{>0})\} \cup \mathcal{U}_v^{>0} \cup (R_w^{>0} \setminus R_v^{>0}).$$
Thus,
$$G_w = v(\mathcal{U}_w^{>0}) = \{-v(a) \mid a \in (R_w^{>0} \setminus R_v^{>0})\} \cup \{0\} \cup \{v(a) \mid a \in (R_w^{>0} \setminus R_v^{>0})\}$$
with
$$G_w^{>0} = \{-v(a) \mid a \in (R_w^{>0} \setminus R_v^{>0})\}$$
and
$$G_w^{<0} = \{v(a) \mid a \in (R_w^{>0} \setminus R_v^{>0})\}.$$
Now the natural valuation v_G satisfies $v_G(g) = v_G(-g)$ for all $g \in G$. Thus,
$$v_G(v(\mathcal{U}_w^{>0})) = v_G(G_w) = v_G(G_w^{<0}) = v_G(\{v(a) \mid a \in (R_w^{>0} \setminus R_v^{>0})\}),$$
as required. □

We turn to the Taylor axiom (T_1).

Lemma 3.16 *If ℓ is the inverse of an exponential f, then the Taylor axiom (T_1) is equivalent to*
$$v(b - \ell(1+b)) > v(b) \qquad \text{for all } b \in I_v. \tag{40}$$

Proof By the v-compatibility, every $a \in I_v$ is of the form $\ell(1+b)$ with $b \in I_v$, and every such $\ell(1+b)$ is in I_v. With $a = \ell(1+b)$, the assertion $v(f(a)-1-a) > v(a)$ is equivalent to $v(b - \ell(1+b)) > v(\ell(1+b))$. But as this implies that $v(b) = v(\ell(1+b))$, it is equivalent to $v(b - \ell(1+b)) > v(b)$. □

This result leads to the following definition: a prelogarithm ℓ will be called a **(T_1)-prelogarithm** if it satisfies condition (40).

Lemma 3.17 *For a* (T_1)*-prelogarithm* ℓ*, the condition* $\ell(1 + I_w) = I_w \cap \mathrm{im}(\ell)$ *holds for all convex valuations* w*.*

Proof Condition (40) implies that $v(b) = v(\ell(1+b))$ for all $b \in I_v$ and therefore, that $\ell(1 + I_w) \subset I_w$ and $\ell(1 + I_v \setminus 1 + I_w) \subset I_v \setminus I_w$ for every convex valuation w. By our general assumption, ℓ is compatible with v, so we have that $\ell(1 + I_v) = I_v \cap \mathrm{im}(\ell) \supset I_w \cap \mathrm{im}(\ell)$. Consequently, $\ell(1 + I_w) = I_w \cap \mathrm{im}(\ell)$. $\square$

By this lemma, a (T_1)-prelogarithm ℓ always satisfies the second condition of (27). So we have proved:

Lemma 3.18 *Let ℓ be a* (GA)*- and* (T_1)*-prelogarithm. Then a convex valuation w is compatible with ℓ if and only if it satisfies* $\ell(K^{>0} \setminus R_w) \subset K^{>0} \setminus R_w$*. Further, the exponential rank $\mathcal{R}_\ell$ is precisely the subset of all $R_w \in \mathcal{R}$ for which $w \neq v$ and w is compatible with ℓ.*

4 The shift map ζ_ℓ

Throughout this section, we assume ℓ to be a (GA)-prelogarithm, compatible with v, but not necessarily (T_1).

Recall that χ_ℓ was defined in Chapter 2, for a logarithm ℓ. We can define $\chi_\ell : G^{<0} \to G^{<0}$ exactly in the same way for a prelogarithm ℓ:

$$\chi_\ell(v(a)) := v(\ell(a)) \qquad \text{for all } a \in K^{>0} \setminus R_v .$$

Recall that this definition does not depend upon the representative a of the value $v(a)$. Let us sketch the argument once more: if $a, b \in K^{>0} \setminus R_v$ such that $v(a) = v(b)$, then $a = bc$ with $c \in \mathcal{U}_w^{>0}$. It follows that $\ell(a) = \ell(bc) = \ell(b) + \ell(c)$, with $\ell(c) \in R_v$. Since $v(\ell(b)) < 0$, we obtain that $v(\ell(a)) = v(\ell(b) + \ell(c)) = \min\{v(\ell(b)), v(\ell(c))\} = v(\ell(b))$.

Clearly χ_ℓ satisfies axiom (C2) and (C3) for contractions, that is, χ_ℓ preserves $\leq$ and contracts archimedean classes. It satisfies axiom (C1), that is, χ_ℓ is surjective *if and only if* the prelogarithm ℓ is surjective. Thus we call χ_ℓ the **precontraction** induced by the prelogarithm ℓ.

We now want to introduce a map ζ_ℓ that the prelogarithm induces on the value set Γ. We will use χ_ℓ and ζ_ℓ to characterize the exponential rank.

- **Definition of the map ζ_ℓ on the value set $v_G(G)$.**

Since every two archimedean equivalent elements of $G^{<0}$ have the same image under χ_ℓ, the map $\zeta_\ell : v_G(G) \to v_G(G)$ given by

$$\zeta_\ell(v_G(g)) := v_G(\chi_\ell(g)) \qquad \text{for all } g \in G^{<0}$$

is well-defined.

By definition of χ_ℓ and ζ_ℓ, we have the following commutative diagram:

4. The shift map ζ_ℓ

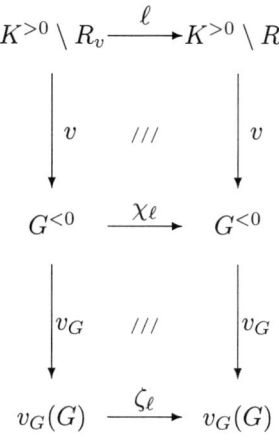

In this diagram, the map v reverses the order $\leq$, and v_G preserves the order $\leq$. Moreover, both are onto. Since also ℓ preserves the order $\leq$ (i.e., ℓ is monotone), we find that:

a) χ_ℓ and ζ_ℓ are monotone,

b) if ℓ is onto, then so are χ_ℓ and ζ_ℓ.

Since χ_ℓ and ζ_ℓ are in general not injective, they may not be strictly monotone.

(If v is the trivial valuation, then $K^{>0} \setminus R_v$, $G^{<0}$ and $v_G(G)$ are empty and χ_ℓ and ζ_ℓ are the empty maps.)

We call ζ_ℓ the **shift map** induced by the prelogarithm ℓ.

Since χ_ℓ preserves $\leq$ and sends archimedean equivalent elements (i.e., elements with equal v_G-value) to the same point, the following holds:

$$\left. \begin{array}{l} v_G(g) = v_G(g') \Rightarrow \chi_\ell(g) = \chi_\ell(g') \\ v_G(g) \geq v_G(g') \Rightarrow \chi_\ell(g) \geq \chi_\ell(g') \end{array} \right\} \quad \text{for all } g, g' \in G^{<0} . \qquad (41)$$

From (33) we infer:

$$g < \chi_\ell(g) \quad \text{for all } g \in G^{<0} . \qquad (42)$$

It follows that $v_G(g) \leq v_G(\chi_\ell(g))$ for all $g \in G^{<0}$. But $v_G(g) = v_G(\chi_\ell(g))$ cannot hold; otherwise (41) would yield that $\chi_\ell(g) = \chi_\ell(\chi_\ell(g))$, in contradiction to (42). So we find:

$$\gamma < \zeta_\ell(\gamma) \quad \text{for all } \gamma \in \Gamma . \qquad (43)$$

Conversely, if ζ_ℓ satisfies (43), then χ_ℓ satisfies (42). For if not, then $g \geq \chi_\ell(g)$ for some $g \in G^{<0}$. But then by (7) we have that $v_G(g) \geq v_G(\chi_\ell(g))$, which contradicts (43) for $\gamma = v_G(g)$.

We call χ_ℓ a centripetal precontraction if it satisfies (42), and ζ_ℓ a **shift to the right** if it satisfies (43). Note that if χ is a centripetal precontraction on G, then even more: $v_G(\chi(g)) > v_G(g)$ for all $g \in G^{<0}$, cf. Lemma A.8, part b).

We have proved the following important lemma:

Lemma 3.19 *Let ℓ be a prelogarithm on K. Then the following are equivalent:*

(i) ℓ is a (GA)-prelogarithm.

(ii) χ_ℓ is a centripetal precontraction.

(iii) ζ_ℓ is a shift to the right.

- **Equivalence relations induced by ℓ, χ_ℓ and ζ_ℓ.**

Remark 3.20 Let φ be any map from a totally ordered set S into itself. By induction, we define $\varphi^1(a) := \varphi(a)$ and $\varphi^{n+1}(a) := \varphi(\varphi^n(a))$.

We define a relation $\sim_\varphi$ on S by setting $a \sim_\varphi a'$ if the convex hulls of the sets $\{a, \varphi^n(a) \mid n \in \mathbb{N}\}$ and $\{a', \varphi^n(a') \mid n \in \mathbb{N}\}$ have a non-empty intersection. This relation is in general not transitive. But if φ is monotone, it is an equivalence relation. In this case, we will say that a and a' are φ-**equivalent** if $a \sim_\varphi a'$. The equivalence classes $[a]_\varphi$ of $\sim_\varphi$ are convex and closed under application of φ. By the convexity, the order of S induces an order on $S/\sim_\varphi$ such that $[a]_\varphi < [b]_\varphi$ if and only if $a' < b'$ for all $a' \in [a]_\varphi$ and $b' \in [b]_\varphi$. On the positive part or the negative part of an ordered abelian group G, the archimedean equivalence relation is obtained by setting $\varphi(a) := 2a$, and v_G is the map $a \mapsto [a]_\varphi$. The order we have introduced on $v_G(G) = v_G(G^{<0})$ is just the one induced by the order of $G^{<0}$.

The maps ℓ, χ_ℓ and ζ_ℓ are monotone. Through the above definition, they induce corresponding equivalence relations on $K^{>0} \setminus R_v$, $G^{<0}$ and Γ respectively. Since we assume ℓ to be a (GA)-prelogarithm, the orientation of these maps is fixed (cf. (32), (42) and (43)). Therefore, we have for $a, a' \in K^{>0} \setminus R_v$ that $a \sim_\ell a'$ if and only if there is some $n \in \mathbb{N}$ such that
$$\ell^n(a) \leq (a') \text{ and } \ell^n(a') \leq a.$$
Similarly, if the exponential f is the inverse of a (GA)-logarithm ℓ, then $a \sim_f a'$ if and only if there is some $n \in \mathbb{N}$ such that
$$f^n(a) \geq a' \text{ and } f^n(a') \geq a.$$
Moreover, the relations $\sim_\ell$ and $\sim_f$ will coincide since
$$f^n(a) \geq a' \text{ if and only if } a \geq \ell^n(a').$$
If $g, g' \in G^{<0}$, then $g \sim_{\chi_\ell} g'$ if and only if there is some $n \in \mathbb{N}$ such that
$$\chi_\ell^n(g) \geq g' \text{ and } \chi_\ell^n(g') \geq g.$$
Similarly, if $\gamma, \gamma' \in \Gamma$, then $\gamma \sim_{\zeta_\ell} \gamma'$ if and only if there is some $n \in \mathbb{N}$ such that
$$\zeta_\ell^n(\gamma) \geq \gamma' \text{ and } \zeta_\ell^n(\gamma') \geq \gamma.$$

Lemma 3.21 *For all $a, a' \in K^{>0} \setminus R_v$, the assertions $a \sim_\ell a'$, $v(a) \sim_{\chi_\ell} v(a')$ and $v_G(v(a)) \sim_{\zeta_\ell} v_G(v(a'))$ are equivalent.*

Proof Set $v(a) = g$ and $v(a') = g'$. Suppose that $g \sim_{\chi_\ell} g'$ and take $n \in \mathbb{N}$ such that $\chi_\ell^n(g) \geq g'$ and $\chi_\ell^n(g') \geq g$. Then $\zeta_\ell^n(v_G(g)) = v_G(\chi_\ell^n(g)) \geq v_G(g')$ and $\zeta_\ell^n(v_G(g')) = v_G(\chi_\ell^n(g')) \geq v_G(g)$. That is, $v_G(g) \sim_{\zeta_\ell} v_G(g')$.

For the converse, suppose the latter and take $n \in \mathbb{N}$ such that $v_G(\chi_\ell^n(g)) = \zeta_\ell^n(v_G(g)) \geq v_G(g')$ and $v_G(\chi_\ell^n(g')) = \zeta_\ell^n(v_G(g')) \geq v_G(g)$. By (41) and (42), this implies that $\chi_\ell^{n+1}(g) \geq \chi_\ell(g') \geq g'$ and $\chi_\ell^{n+1}(g') \geq \chi_\ell(g) \geq g$. That is, $g \sim_{\chi_\ell} g'$.

Now it remains to show that $a \sim_\ell a'$ and $v(a) \sim_{\chi_\ell} v(a')$ are equivalent. Suppose that $a \sim_\ell a'$ and take $n \in \mathbb{N}$ such that $\ell^n(a) \leq a'$ and $\ell^n(a') \leq a$. Then $\chi_\ell^n(v(a)) = v(\ell^n(a)) \geq v(a')$ and $\chi_\ell^n(v(a')) = v(\ell^n(a')) \geq v(a)$. That is, $v(a) \sim_{\chi_\ell} v(a')$. In particular, we obtain that $v(a) \sim_{\chi_\ell} v(a')$.

For the converse, assume the latter and take $n \in \mathbb{N}$ such that $\chi_\ell^n(v(a)) \geq v(a')$ and $\chi_\ell^n(v(a')) \geq v(a)$. By (42), we obtain that $v(\ell^{n+1}(a)) = \chi_\ell^{n+1}(v(a)) > \chi_\ell^n(v(a)) \geq v(a')$ and $v(\ell^{n+1}(a')) = \chi_\ell^{n+1}(v(a')) > \chi_\ell^n(v(a')) \geq v(a)$. Consequently, $\ell^{n+1}(a) < a'$ and $\ell^{n+1}(a') < a$. That is, $a \sim_\ell a'$. $\square$

Below, denote the $\sim_\ell$-equivalence class of a by $[a]_\ell$, and call it the **exponential class** of a. Similarly for the $\sim_{\chi_\ell}$- and $\sim_{\zeta_\ell}$- equivalence classes.

Corollary 3.22 *(i) The map $[a]_\ell \mapsto [v(a)]_{\chi_\ell}$ is an order reversing bijection between $(K^{>0} \setminus R_v)/\sim_\ell$ and $G^{<0}/\sim_{\chi_\ell}$.*
(ii) The map $[g]_{\chi_\ell} \mapsto [v_G(g)]_{\zeta_\ell}$ is an order preserving bijection between $G^{<0}/\sim_{\chi_\ell}$ and $\Gamma/\sim_{\zeta_\ell}$.

Suppose that $\sim_1$ and $\sim_2$ are two equivalence relations defined on the same set. We say that $\sim_1$ is coarser than $\sim_2$ if $\sim_2$-equivalence implies $\sim_1$-equivalence.

4. The shift map ζ_ℓ

Lemma 3.23 *The equivalence relation $\sim_\ell$ is coarser than the archimedean equivalence relations with respect to addition and multiplication on $K^{>0} \setminus R_v$, and $\sim_{\chi_\ell}$ is coarser than the archimedean equivalence relation on $G^{<0}$. In other words, the equivalence classes of $\sim_\ell$ are closed under addition and multiplication, and those of $\sim_{\chi_\ell}$ are closed under addition.*

Proof Assume that $a, a' \in K^{>0} \setminus R_v$ such that $a < a' < na$. Since ℓ is a (GA)-prelogarithm, we have $v(a) = v(na) < v(\ell(na))$ and thus, $\ell(a') < \ell(na) < a$. This proves that archimedean equivalence with respect to addition implies ℓ-equivalence. Now if $a < a' < a^n$, then $\ell(a) < \ell(a') < n\ell(a)$, and by what we have already shown, $\ell(a) \sim_\ell \ell(a')$. Since $\ell(b) \sim_\ell b$ for every $b \in K^{>0} \setminus R_v$, it follows that $a \sim_\ell \ell(a) \sim_\ell \ell(a') \sim_\ell a'$. This proves that archimedean equivalence with respect to multiplication implies ℓ-equivalence. In view of Lemma 3.21 and the fact that $v(ab) = v(a) + v(b)$, this result also yields our assertion about $\sim_{\chi_\ell}$. □

Recall that if K is non-archimedean and ℓ is surjective, then G is divisible and Γ is dense without endpoints (cf. Lemma 1.23). This holds also for the equivalence classes:

Proposition 3.24 *Assume that K is non-archimedean and ℓ is surjective. Then with the induced ordering, every χ_ℓ-equivalence class and every ζ_ℓ-equivalence class is dense without endpoints.*

Proof If ℓ is surjective, then so are χ_ℓ and ζ_ℓ. In view of (42) and (43), this yields that their equivalence classes have no endpoints. Since the χ_ℓ-equivalence classes and the ζ_ℓ-equivalence classes are convex subsets of G and Γ, it follows that they are also dense. □

Now we are able to give the promised characterization (Theorem 3.25 and Corollary 3.27 below). Recall that
$$\Gamma_w = v_G(G_w) = v_G(G_w^{<0})$$
for every convex subgroup G_w of G. Since $G_w^{<0}$ is a final segment of $G^{<0}$ and v_G preserves $\leq$ on $G_w^{<0}$, Γ_w is a final segment of Γ. We note that
$$a \in R_w \Leftrightarrow v(a) \in G_w \Leftrightarrow v_G(v(a)) \in \Gamma_w \quad \text{for all } a \in K^{>0} \setminus R_v \qquad (44)$$
(the second equivalence and the implication $v(a) \in G_w \Rightarrow a \in R_w$ hold more generally for all $a \in K$). Indeed, the implication $v(a) \in G_w \Rightarrow v_G(v(a)) \in \Gamma_w$ holds by definition of Γ_w. The converse holds since the convex subgroup G_w is closed under archimedean equivalence. Further, we note that $v(a) = v(b)$ implies that $w(a) = w(b)$. Hence, $v(a) \in G_w$ implies that $w(a) = 0$, whence $a \in R_w$. For the converse, observe that every $a \in K^{>0} \setminus R_v$ satisfies $v(a) < 0$ and thus, $w(a) \leq 0$. If in addition $a \in R_w$, then $a \in \mathcal{U}_w^{>0}$ and consequently, $v(a) \in G_w$.

The following theorem gives criteria for R_w to belong to the exponential rank:

Theorem 3.25 *Let ℓ be a (GA)-prelogarithm. Then the following assertions are equivalent:*
a) $\ell(\mathcal{U}_w^{>0}) = R_w \cap \text{im}(\ell)$ *b) $\ell(K^{>0} \setminus R_w) \subset K^{>0} \setminus R_w$*
c) $G^{<0} \setminus G_w$ is closed under χ_ℓ   *d) $G_w^{<0}$ is closed under χ_ℓ-equivalence*
e) $\Gamma \setminus \Gamma_w$ is closed under ζ_ℓ   *f) Γ_w is closed under ζ_ℓ-equivalence.*

If ℓ is surjective, and its inverse is the exponential $f = \ell^{-1}$, then these conditions are also equivalent to

g) $a \in R_w \Rightarrow f(a) \in R_w$ for all $a \in K^{>0} \setminus R_v$
h) $v(a) \in G_w \Rightarrow v(f(a)) \in G_w$ for all $a \in K^{>0} \setminus R_v$
i) $v_G(v(a)) \in \Gamma_w \Rightarrow v_G(v(f(a))) \in \Gamma_w$ for all $a \in K^{>0} \setminus R_v$.

Proof a) $\Leftrightarrow$ b): This was already shown in the last section.
b) $\Leftrightarrow$ c): We know from the last section that condition b) is equivalent to (30). But $w(a) < 0$ is equivalent to $v(a) < G_w$, and $w(\ell(a)) < 0$ is equivalent to $\chi_\ell v(a) = v(\ell(a)) < G_w$. Thus, (30) is equivalent to condition c).
c) $\Rightarrow$ d): Suppose that $g \in G_w^{<0}$ and that $g' \in G^{<0}$ with $g \sim_{\chi_\ell} g'$. Take $n \in \mathbb{N}$ such that $\chi_\ell^n(g') \geq g$. Since $G_w^{<0}$ is a final segment of $G^{<0}$, it follows that $\chi_\ell^n(g') \in G_w^{<0}$. Since $G^{<0} \setminus G_w$ is assumed to be closed under χ_ℓ, this implies that $g' \in G_w^{<0}$.
d) $\Rightarrow$ c): Take $g \in G^{<0} \setminus G_w$. Since $\chi_\ell(g) \sim_{\chi_\ell} g$ and $G_w^{<0}$ is assumed to be closed under χ_ℓ-equivalence, we find that $\chi_\ell(g) \in G^{<0} \setminus G_w$.
c) $\Leftrightarrow$ e): Follows directly from the definition of ζ_ℓ.
e) $\Leftrightarrow$ f): Similar to the proof of c) $\Leftrightarrow$ d).

Now suppose that $\ell = f^{-1}$.

d) $\Rightarrow$ h): Follows from the fact that $v(a) = v(\ell(f(a))) = \chi_\ell(v(f(a)))$ is χ_ℓ-equivalent to $v(f(a))$.
h) $\Rightarrow$ d): Suppose that $g \in G_w^{<0}$ and that $g' \in G^{<0}$ with $g \sim_{\chi_\ell} g'$. Choose $a' \in K^{>0}$ such that $g' = v(a')$. Take $n \in \mathbb{N}$ such that $\chi_\ell^n(g') \geq g$. Since $G_w^{<0}$ is a final segment of $G^{<0}$, it follows that $v(\ell^n(a')) = \chi_\ell^n(g') \in G_w^{<0}$. By n-fold application of h), we find that $g' = v(a') = v(f^n \ell^n a') \in G_w$.
g) $\Leftrightarrow$ h) $\Leftrightarrow$ i): Follows from (44). $\square$

We add the following lemma which we shall use in the next section:

Lemma 3.26 $\ell(K^{>0} \setminus R_w) \subset K^{>0} \setminus R_w$ if and only if $R_w^{>0} \setminus R_v^{>0}$ is closed under ℓ − equivalence.

Proof Indeed, assume the first condition holds. Let $a \in (R_w^{>0} \setminus R_v^{>0})$ and $b \in K^{>0} \setminus R_v$ such that $b \sim_\ell a$. Let $n \in \mathbb{N}$ be such that $\ell^n(b) \leq a$. We show that $b \in (R_w^{>0} \setminus R_v^{>0})$. If $b \notin R_w$, then by assumption $\ell^n(b) \in K^{>0} \setminus R_w$. But this contradicts the convexity of R_w.

Conversely, assume that the second condition holds. Let $a \in K^{>0} \setminus R_w$, we show that $\ell(a) \in K^{>0} \setminus R_w$. Now a is positive infinite and ℓ is v-compatible implies that $\ell(a)$ is positive infinite as well, that is, $v(\ell(a)) < 0$. So $\ell(a) \notin R_v$. Thus if $\ell(a) \in R_w$ then $\ell(a) \in (R_w^{>0} \setminus R_v^{>0})$. Clearly $a \sim_\ell \ell(a)$. Thus by assumption $a \in (R_w^{>0} \setminus R_v^{>0})$, which is a contradiction. $\square$

Corollary 3.27 Let f be a (GAT_1)-exponential. Then a convex valuation w is compatible with f if and only if for every $a \in K$,

$$v(a) \in G_w \Rightarrow v(f(a)) \in G_w. \tag{45}$$

Proof Take an exponential f which satisfies (T_1) and (GA). Then $\ell = f^{-1}$ is a (GAT_1)-logarithm. Suppose first that w is compatible with f. By our remark following (44) and by the first condition of (CE),

$$v(a) \in G_w \Rightarrow a \in R_w \Rightarrow f(a) \in \mathcal{U}_w^{>0} \Rightarrow w(f(a)) = 0 \Rightarrow v(f(a)) \in G_w.$$

Now suppose that (45) holds. Then in particular, condition h) of the foregoing theorem holds, which proves that w is compatible with f. $\square$

5 Characterization of the exponential and the principal exponential rank

We shall now use the full strength of Theorem 3.25. We let ℓ be a (GA)-prelogarithm on K. As always, we denote by G the value group $v(K)$ and by Γ the value set $v_G(v(K))$. If $R_w \neq R_v$ is a convex valuation subring of K, we say that R_w is ℓ-**closed** if $R_w^{>0} \setminus R_v^{>0}$ is closed under ℓ-equivalence. Recall that this is equivalent to $\ell(\mathcal{U}_w^{>0}) = R_w \cap \mathrm{im}(\ell)$, by Theorem 3.25 and Lemma 3.26. By definition, it implies that $R_w \in \mathcal{R}_\ell$.

We say that R_w is ℓ-**principal** if there is $a \in K^{>0} \setminus R_v$ such that R_w is the smallest ℓ-closed convex subring containing a (so that actually $a \in R_w^{>0} \setminus R_v^{>0}$). In that case we say that R_w is ℓ-**principal generated by** a. We define the **principal exponential rank** of (K, ℓ) to be the subset $\mathcal{R}_\ell^{\mathrm{pr}}$ of $\mathcal{R}_\ell$ consisting of all $R_w \in \mathcal{R}_\ell$ which are ℓ-principal.

We want to analyze the principal exponential rank of (K, ℓ), as we did for the principal rank of K.

A convex subgroup $G_w \neq 0$ of G is said to be χ_ℓ-**closed** if $G_w^{\leq 0}$ is closed under χ_ℓ-equivalence. It is χ_ℓ-**principal** if there is $g \in G^{<0}$ such that G_w is the smallest χ_ℓ-closed convex subgroup containing g. In that case we say that G_w is χ_ℓ-**principal generated by** g.

A non-empty final segment Γ_w of Γ is said to be ζ_ℓ-**closed** if it is closed under ζ_ℓ-equivalence. It is ζ_ℓ-**principal** if there is $\gamma \in \Gamma$ such that Γ_w is the smallest ζ_ℓ-closed final segment containing γ. In that case we say that Γ_w is ζ_ℓ-**principal generated by** γ.

By the definitions of the ℓ-, χ_ℓ- and ζ_ℓ-equivalence relations, and -closedness we have:

Lemma 3.28 *(i) If R_w is an ℓ-closed valuation subring, then $(R_w^{>0} \setminus R_v^{>0})/\sim_\ell$ is an initial segment of $(K^{>0} \setminus R_v)/\sim_\ell$.*

(ii) If G_w is a χ_ℓ-closed convex subgroup, then $G_w^{\leq 0}/\sim_{\chi_\ell}$ is a final segment of $G^{<0}/\sim_{\chi_\ell}$.

(iii) If Γ_w is a ζ_ℓ-closed final segment, then $\Gamma_w/\sim_{\zeta_\ell}$ is a final segment of $\Gamma/\sim_{\zeta_\ell}$.

The following lemma can be immediately deduced from Theorem 3.25 and Lemma 3.26.

Lemma 3.29 *Let $w \neq v$ be a convex valuation. Then R_w is ℓ-closed if and only if $G_w = v(\mathcal{U}_w^{>0})$ is χ_ℓ-closed, if and only if $\Gamma_w = v_G(G_w)$ is ζ_ℓ-closed.*

We can now prove the desired characterization of the exponential rank. Let $(\Gamma/\sim_{\zeta_\ell})^{\mathrm{fs}}$ denote the set of non-empty final segments of $\Gamma/\sim_{\zeta_\ell}$, totally ordered by inclusion.

Theorem 3.30 (Characterization of the exponential rank) *Let ℓ be a (GA)-prelogarithm. Then*

$$\varepsilon : R_w \mapsto \{[v_G(v(a))]_{\zeta_\ell} \mid a \in \mathcal{U}_w^{>0}\}$$

is an order preserving bijection from the exponential rank $\mathcal{R}_\ell$ onto $(\Gamma/\sim_{\zeta_\ell})^{\mathrm{fs}}$.

Proof First note that ε is well-defined by Lemma 3.28. Now note that

$$\varepsilon(R_w) = \{[v_G(v(a))]_{\zeta_\ell} \mid a \in R_w^{>0} \setminus R_v^{>0}\}$$

by Lemma 3.15. Note also that the map

$$[a]_\ell \mapsto [v_G(v(a))]_{\zeta_\ell}$$

is an order reversing bijection from $(K^{>0} \setminus R_v)/\sim_\ell$ onto $\Gamma/\sim_{\zeta_\ell}$ by Corollary 3.22. Let $((K^{>0} \setminus R_v)/\sim_\ell)^{\text{is}}$ denote the set of non-empty initial segments of $(K^{>0}\setminus R_v)/\sim_\ell$, totally ordered by inclusion. Then by our considerations above, we see that the map

$$\varepsilon_1 : \mathcal{I} \mapsto \{[v_G(v(a))]_{\zeta_\ell} \mid [a]_\ell \in \mathcal{I}\}$$

is an order preserving bijection from $((K^{>0} \setminus R_v)/\sim_\ell)^{\text{is}}$ onto $(\Gamma/\sim_{\zeta_\ell})^{\text{fs}}$. Thus to establish the assertion of the theorem, it remains to show that the map

$$\varepsilon_2 : R_w \mapsto \{[a]_\ell \mid a \in R_w^{>0} \setminus R_v^{>0}\}$$

is an order preserving bijection from the exponential rank $\mathcal{R}_\ell$ onto $((K^{>0}\setminus R_v)/\sim_\ell)^{\text{is}}$. Then clearly $\varepsilon = \varepsilon_1 \circ \varepsilon_2$ will be the compositum of two order preserving bijections, and thus is an order preserving bijection.

Clearly, ε_2 is order preserving and one-to-one, since the rings are ℓ-closed. The main effort is to show that ε_2 is onto. To relax the notation, set

$$\mathcal{I}_w = \{[a]_\ell \mid a \in R_w^{>0} \setminus R_v^{>0}\}.$$

We want to show that given $\mathcal{I} \in ((K^{>0} \setminus R_v)/\sim_\ell)^{\text{is}}$, there is an ℓ-closed convex valuation subring R_w such that $\mathcal{I}_w = \mathcal{I}$. Given $\mathcal{I}$, let $(\bigcup \mathcal{I})$ denote the set theoretic union of the elements of $\mathcal{I}$ and $-(\bigcup \mathcal{I})$ the set of additive inverses. Set

$$R_w = -\left(\bigcup \mathcal{I}\right) \cup R_v \cup \left(\bigcup \mathcal{I}\right).$$

We claim that R_w is the required ring. Clearly R_w is ℓ-closed, and $\mathcal{I}_w = \mathcal{I}$. Further R_w is convex (by its construction), and strictly contains R_v. We now have to show that R_w is a ring.

Let $a, b \in R_w$, we show $a + b \in R_w$. We may assume a or $b \notin R_v$ (otherwise there is nothing to prove, since R_v is a subring). If a, b are both positive, then by (12) $v(a+b) = \min\{v(a), v(b)\}$. Thus $a+b$ is archimedean equivalent to either a or b, hence $a + b$ is positive infinite. Thus by Lemma 3.23 $a + b$ is ℓ-equivalent equivalent to either a or b and we are done. In all other cases, we have $|a + b| \leq |a| + |b|$ and we are done by convexity and what we have just shown. Let $a, b \in R_w$, we show $ab \in R_w$. We may assume that a and b are non-zero. Since obviously R_w is closed under additive inverses, it suffices to show that $|ab| \in R_w$. Again, assume that a or $b \notin R_v$ and first assume that both $|a| > 1$ and $|b| > 1$. So we can apply (12) to the natural valuation $v\dot{}$. Thus $v\dot{}(|ab|) = \min\{v\dot{}(|a|), v\dot{}(|b|)\}$. Thus $|ab|$ is multiplicatively equivalent to either $|a|$ or $|b|$, hence $|ab|$ is positive infinite. Thus by Lemma 3.23, $|ab|$ is ℓ-equivalent equivalent to either $|a|$ or $|b|$ and we are done. Now if $|b| < 1$, then $|ab| = |a||b| \leq |a|$ and we are done by convexity. $\square$

The following lemmas can be immediately deduced from Theorem 3.25.

Lemma 3.31 *Let ℓ be a (GA)-prelogarithm and $w \neq v$ be a convex valuation. Then R_w is ℓ-principal generated by a if and only if $G_w = v(\mathcal{U}_w^{>0})$ is χ_ℓ-principal generated by $v(a)$, if and only if $\Gamma_w = v_G(G_w)$ is ζ_ℓ-principal generated by $v_G(v(a))$.*

Lemma 3.32 *Let ℓ be a (GA)-prelogarithm on K. Let Γ_w be a ζ_ℓ-closed final segment of Γ. Then Γ_w is ζ_ℓ-principal generated by γ if and only if the ζ_ℓ-equivalence class $[\gamma]_{\zeta_\ell}$ is an initial segment in Γ_w.*

By the last lemma it follows that Γ_w is ζ_ℓ-principal if and only if the final segment $\Gamma_w/\sim_{\zeta_\ell}$ of $\Gamma/\sim_{\zeta_\ell}$ has a least element.

5. Characterization of the exponential and the principal exponential rank

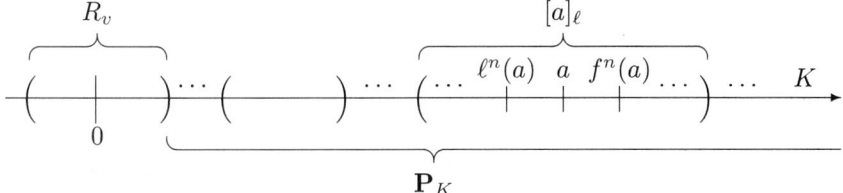

The above lemmas together with Lemma 3.22 prove the exponential analogue of Theorem 3.9.

Theorem 3.33 *Let ℓ be a (GA)-prelogarithm and $w \neq v$ be a convex valuation on K. Set $G_w = v(\mathcal{U}_w^{>0})$ and $\Gamma_w = v_G(G_w)$. Then the following are equivalent:*
(i) R_w is ℓ-principal generated by a.
(ii) The class $[a]_\ell$ is the largest amongst all $\sim_\ell$- classes of R_w (this is equivalent to $[a]_\ell$ is a final segment in R_w).
(iii) The class $[v(a)]_{\chi_\ell}$ is the smallest amongst all $\sim_{\chi_\ell}$-classes of $G_w^{<0}$ (this is equivalent to $[v(a)]_{\chi_\ell}$ is an initial segment in G_w).
(iv) The class $[v_G(v(a))]_{\zeta_\ell}$ is the smallest amongst all $\sim_{\zeta_\ell}$-classes of Γ_w (this is equivalent to $[v_G(v(a))]_{\zeta_\ell}$ is an initial segment in Γ_w).

Corollary 3.34 (Characterization of the principal exponential rank)
Let ℓ be a (GA)-prelogarithm on K. Then

$$R_w \mapsto \min\{[v_G(v(a))]_{\zeta_\ell} \mid a \in R_w^{>0} \setminus R_v^{>0}\}$$

is an order reversing bijection from the principal exponential rank onto $\Gamma/\sim_{\zeta_\ell}$.

Thus if $(\Gamma/\sim_{\zeta_f})^*$ denotes the set $\Gamma/\sim_{\zeta_f}$ endowed with the reversed ordering we get the exponential analogue to Lemma 3.8:

Corollary 3.35 *Let ℓ be a (GA)-prelogarithm on K. Then the principal exponential rank of (K, ℓ) is (isomorphic to) $(\Gamma/\sim_{\zeta_\ell})^*$.*

To summarize: $\varepsilon(R_w)$ contains a smallest element if and only if Γ_w admits some $\sim_{\zeta_\ell}$-equivalence class as an initial segment, or equivalently, $G_w^{\leq 0}$ admits some χ_ℓ-equivalence class as an initial segment. But this does not mean that G_w is principal; the following corollary shows the contrary.

Corollary 3.36 *If ℓ is surjective, then the intersection of the principal rank and the exponential rank is empty. In particular, the value group of a non-archimedean exponential field is never principal (as its own convex subgroup).*

Proof If R_w belongs to the exponential rank, then $v_G(G_w)$ is closed under ζ_ℓ-equivalence. If ℓ is surjective, then Theorem 3.24 shows that ζ_ℓ-equivalence classes have no smallest element; hence also $v_G(G_w)$ has no smallest element. The second assertion follows from the first, taking w to be the trivial valuation. □

Now let f be a (GA)-exponential on K, and $w \neq v$ a convex valuation. We say that R_w is f-closed if $f(R_w) = \mathcal{U}_w^{>0}$. Set $\ell = f^{-1}$. We define the principal exponential rank of (K, f) to be that of (K, ℓ). The following corollary follows from the above results, together with Lemma 3.17.

Corollary 3.37 *Let f be a (GAT_1)-exponential on K, and $w \neq v$ a convex valuation. Set $\ell = f^{-1}$. Then R_w is f-closed if and only if w and f are compatible. Thus the exponential rank of (K, f) is the chain of f-compatible convex* valuation rings (containing R_v properly). It is isomorphic to the chain $(\Gamma/\sim_{\zeta_\ell})^{\text{fs}}$. The principal exponential rank of (K,f) is isomorphic to $(\Gamma/\sim_{\zeta_\ell})^*$.

Exponential rank and principal exponential rank describe the growth of an exponential f in comparison to the size of K. To determine this growth, we can look at sequences generated by repeated application of f. We have the exponential analogue of Theorem 3.10:

Theorem 3.38 *Let $w \neq v$ be a convex valuation on the non-archimedean ordered field $(K,<)$. Assume in addition that f is a (GAT_1)-exponential and that f is compatible with w. Then R_w lies in the principal exponential rank if and only if for some b in the residue field Kw, the sequence $((fw)^n(b))_{n\in\mathbb{N}}$ is cofinal in Kw. The principal exponential rank is a singleton (namely, $\mathcal{R}_f^{\text{pr}} = \{K\}$) if and only if $(f^n(a))_{n\in\mathbb{N}}$ is cofinal in K for every positive infinite element $a \in K$.*

Proof Take any $a \in K^{>0} \setminus R_v$ and $R_w \in \mathcal{R}_f$, $w \neq v$. From the above results it follows that the sequence $(f^n(a))_{n\in\mathbb{N}}$ is cofinal in the $\sim_f$-equivalence class $[a]_f$ of a, the sequence $(v(f^n(a)))_{n\in\mathbb{N}}$ is coinitial in the $\sim_{\chi_\ell}$-equivalence class $[v(a)]_{\chi_\ell}$ of $v(a)$, and the sequence $(v_G(v(f^n(a))))_{n\in\mathbb{N}}$ is coinitial in the $\sim_{\zeta_\ell}$-equivalence class $[v_G(v(a))]_{\zeta_\ell}$ of $v_G(v(a))$. Hence, the sequence $(f^n a)_{n\in\mathbb{N}}$ is cofinal in R_w if and only if $[v(a)]_{\chi_\ell}$ is an initial segment of G_w, and this is the case if and only if $[v_G(v(a))]_{\zeta_\ell}$ is an initial segment of Γ_w. This in turn holds if and only if $[v_G(v(a))]_{\zeta_\ell}$ is the smallest element in $\varepsilon(R_w)$.

On the other hand, the sequence $(f^n(a))_{n\in\mathbb{N}}$ is cofinal in R_w if and only if the sequence $((fw)^n(aw))_{n\in\mathbb{N}}$ is cofinal in Kw. This follows from $(f^n(a))w = (fw)^n(aw)$ and the fact that the residue map $a \mapsto aw$ induces a $\leq$-preserving map from the set $R_w^{>0} \setminus R_v^{>0}$ onto the positive infinite elements of Kw.

If for every $a \in K^{>0} \setminus R_v$ the sequence $(f^n(a))_{n\in\mathbb{N}}$ is cofinal, then this means that for every such a the class $[v_G(v(a))]_{\zeta_\ell}$ is the same, and vice versa. This in turn means that $\Gamma/\sim_{\zeta_\ell}$ is a singleton, i.e., the principal exponential rank is a singleton. $\square$

Remark 3.39 Finally, let us mention that the characterizations of the exponential rank and the principal exponential rank through equivalence relations is quite natural. Indeed let K be an ordered field and $\Gamma = v_G(v(K))$. We can repeat the procedures described in Section 4, but working with multiplication in place of the exponential. More precisely, we consider the *squaring* function:

$$\varphi : K^{>0} \setminus R_v \to K^{>0} \setminus R_v$$

defined by

$$a \mapsto a^2.$$

Then the induced map $\zeta_\varphi : \Gamma \to \Gamma$ is just the identity map. Thus the induced equivalence relation $\sim_{\zeta_\varphi}$ is the trivial one, and indeed, $\Gamma/\sim_{\zeta_\varphi} \simeq \Gamma$ gives the principal rank of K.

CHAPTER 4

Construction of exponential fields

Let k be an ordered field and G a non-trivial ordered abelian group. Then the power series field $k((G))$ admits at least one non-archimedean order. Further, $k((G))$ is real closed if and only if k is real closed and G is divisible (cf. [PR1]). Moreover, every real closed field embeds in such a power series field (cf. [KA]). This provides a very simple and elegant method of constructing non-archimedean ordered real closed fields. In this chapter, we extend those power series constructions to the exponential case, in order to provide large exponential fields.

The main problem will be to define a logarithm on the multiplicative group of positive elements $k((G))$. We show that indeed one can *always* define a *prelogarithm* if G is divisible (Corollary 4.11). On the other hand, we prove that a prelogarithm can *never* be surjective (Theorem 4.2). Theorem 4.2 shows that the construction method for real closed fields described above is not available for exponential fields. For an even stronger version of Theorem 4.2, see Theorem 4.6.

However, starting with the prelogarithm, we shall develop a procedure of making it surjective on a countable union of power series fields. The exponential fields obtained by this procedure have interesting properties, which shall be studied in detail in the next chapter.

1 w-Logarithmic cross-sections

In this chapter, $(K, <)$ will always be an ordered field. We let v denote the natural valuation on its additive group $(K, +, 0, <)$, and $v(K)$ its value group. We shall always assume that K is root closed for positive elements.

Let w be a convex valuation on K. As was shown for the case $w = v$ in Chapter 1, we have the analogous w-additive and multiplicative lexicographic decompositions:

$$(K, +, 0, <) \simeq \mathbf{A}_w \amalg (Kw, +, 0, <) \amalg (I_w, +, 0, <) \tag{46}$$

where $\mathbf{A}_w$ is an arbitrary group complement of R_w in $(K, +)$, and

$$(K^{>0}, \cdot, 1, <) \simeq \mathbf{B}_w \amalg (Kw^{>0}, \cdot, 1, <) \amalg (1 + I_w, \cdot, 1, <) \tag{47}$$

where $\mathbf{B}_w$ is an arbitrary group complement of $\mathcal{U}_w^{>0}$ in $(K^{>0}, \cdot)$. Endowed with the restriction of the order, $\mathbf{A}_w$ and $\mathbf{B}_w$ are unique up to isomorphism. In view of (CO) and the fact that $w(-a) = wa$, the map

$$(K^{>0}, \cdot, 1, <) \to (w(K), +, 0, <), \qquad a \mapsto -w(a) = w(a^{-1}) \tag{48}$$

is a surjective group homomorphism preserving $\leq$, with kernel $\mathcal{U}_w^{>0}$. We find that every complement $\mathbf{B}_w$ is isomorphic to $(w(K), +, 0, <)$ through the map $-w$.

Let ℓ be a w-compatible prelogarithm. Then ℓ decomposes into three embeddings of ordered groups:

$$\begin{aligned}
\ell_R^w &: (1+I_w,\cdot,1,<) \to (I_w,+,0,<) \\
\ell w &: (Kw^{>0},\cdot,1,<) \to (Kw,+,0,<) \\
\ell_L^w &: \mathbf{B}_w \to \mathbf{A}_w.
\end{aligned}$$

Conversely, in view of (46) and (47), such embeddings ℓ_R^w, ℓw and ℓ_L^w can be put together to obtain a prelogarithm which is compatible with w. We call ℓ_L^w a **w-left prelogarithm** and ℓ_R^w a **w-right prelogarithm**. ℓw is a prelogarithm on the residue field Kw, and ℓ can be seen as a *lifting* of ℓw. Thus, the liftings of ℓw to K are in one-to-one correspondence to the pairs (ℓ_L^w, ℓ_R^w) of w- left and w-right prelogarithms. The set of all w-right prelogarithms is identical to the set of all order preserving embeddings of $(1+I_w,\cdot)$ in $(I_w,+)$; we will denote it by o-Emb$((1+I_w,\cdot),(I_w,+))$.

Through the isomorphism (48), every embedding

$$h: (w(K),+,0,<) \to \mathbf{A}_w$$

gives rise to a w-left prelogarithm $h \circ -w$. Conversely, given a w-left prelogarithm ℓ_L^w, the map

$$h_\ell^w := \ell_L^w \circ (-w)^{-1}$$

is such an embedding (here, $(-w)^{-1}$ is an isomorphism from $w(K)$ onto $\mathbf{B}_w$). Note that h is surjective if and only if ℓ_L^w is. This motivates the following definition. A **w-logarithmic cross-section** of an ordered field $(K,<)$ with respect to a convex valuation w is an order preserving *embedding* h of $w(K)$ in an additive group complement of the valuation ring. Equivalently, h is an embedding of $w(K)$ in $(K,+,0,<)$ such that $h(w(K)) \cap R_w = \{0\}$. The latter is equivalent to that h satisfies $w(h(g)) < 0$ for all $g \in w(K)$. Note that this is an exponential analogue of the classical notion of a cross-section for a valued field, which is of fundamental importance in the theory of valued fields. Every real closed field K admits a **cross-section** with respect to a convex valuation w, that is, an embedding s of the value group $w(K)$ in the multiplicative group $K^\times$ such that $w(s(\alpha)) = \alpha$ for all $\alpha \in w(K)$ (cf. Chapter 6). However, it is not true in general that real closed fields admit logarithmic cross-sections.

We say that h is a **surjective w-logarithmic cross-section** if $h(w(K))$ is a complement of the valuation ring.

Thus there is a one-to-one correspondence between w-left logarithms and surjective w-logarithmic cross-sections.

Recall that we have already encountered surjective v-logarithmic cross-sections in Chapters 1 and 2. These were the isomorphisms h_ℓ induced by v-compatible logarithms ℓ (cf. Remark 1.20). **If $w = v$ we will continue to write h_ℓ instead of h_ℓ^v.**

We will denote by o-Emb$(w(K),(K,+)\setminus R_w)$ the set of all w-logarithmic cross-sections. Further, we denote the set of all prelogarithms of K by $\mathbf{L}_K$, and $\mathbf{L}_K^w$ shall be the subset of those logarithms which are compatible with w.

The following description of the set of all liftings of an order through a place is well-known. If we denote by $\mathbf{X}_K$ the set of all orderings on K, and by $\mathbf{X}_K^w$ the subset of all orderings which are compatible with w, then there is a bijection

$$\mathbf{X}_K^w \to \mathrm{Hom}\,(v(K)/2v(K),\{-1,1\}) \times \mathbf{X}_{Kw}$$

(cf. [LAM1], Theorem 5.3). In the same spirit, we describe the set of all liftings of a prelogarithm:

Theorem 4.1 *The map*

$$\mathbf{L}_K^w \to \text{o-Emb}(w(K), (K, +) \setminus R_w) \times \mathbf{L}_{Kw} \times \text{o-Emb}((1 + I_w, \cdot), (I_w, +))$$
$$\ell \mapsto (h_\ell^w, \ell w, \ell_R^w) \tag{49}$$

is a bijection, and the following holds:
a) ℓ *is surjective if and only if* h_ℓ^w, ℓw *and* ℓ_R^w *are,*
b) *if* w' *is a convex valuation finer than* w, *then* ℓ *is compatible with* w' *if and only if* ℓw *is compatible with the induced valuation* w'/w.

Our goal in the next section is to show that power series fields *always* admit logarithmic cross-sections, but *never* surjective ones.

2 A combinatorial result and its consequences

We shall prove:

Theorem 4.2 *Let k be an ordered field and G a non-trivial ordered abelian group. Let $<$ be any order on $K = k((G))$. Then $(K, <)$ does not admit any $v_{\min}$-compatible exponential. If $(k, <)$ is archimedean, then $(K, <)$ admits no exponential at all.*

The key to our result is the fact that every group complement of the valuation ring in $K = k((G))$ is a lexicographic product of ordered abelian groups. Let us recall the definition of lexicographic products of totally ordered sets.

Let Γ and Δ_γ, $\gamma \in \Gamma$ be totally ordered sets. For every $\gamma \in \Gamma$, we fix a distinguished element $0 \in \Delta_\gamma$. The **support** of $a = (\delta_\gamma)_{\gamma \in \Gamma} \in \coprod_{\gamma \in \Gamma} \Delta_\gamma$, denoted by support($a$), is the set of all $\gamma \in \Gamma$ for which $\delta_\gamma \neq 0$. As a set, we define $\mathbf{H}_{\gamma \in \Gamma} \Delta_\gamma$ to consist of all $(\delta_\gamma)_{\gamma \in \Gamma}$ with well ordered support. The lexicographic order on $\mathbf{H}_{\gamma \in \Gamma} \Delta_\gamma$ is introduced as follows. Given a and $b = (\delta'_\gamma)_{\gamma \in \Gamma} \in \mathbf{H}_{\gamma \in \Gamma} \Delta_\gamma$, observe that support($a$) $\cup$ support(b) is well ordered. Let γ_0 be the least of all elements $\gamma \in$ support(a) $\cup$ support(b) for which $\delta_\gamma \neq \delta'_\gamma$. We set $a < b :\Leftrightarrow \delta_{\gamma_0} < \delta'_{\gamma_0}$. Then $(\mathbf{H}_{\gamma \in \Gamma} \Delta_\gamma, <)$ is a totally ordered set, the lexicographic product of the ordered sets Δ_γ. If $\Delta_\gamma = \Delta$ for all $\gamma \in \Gamma$ then we write Δ^Γ for their lexicographic product; it consists of all maps from Γ to Δ with well ordered support.

If all Δ_γ are ordered abelian groups, then we can take the distinguished elements 0 to be the neutral elements of the groups Δ_γ. Defining addition on $\mathbf{H}_{\gamma \in \Gamma} \Delta_\gamma$ componentwise, we obtain an ordered abelian group $(\mathbf{H}_{\gamma \in \Gamma} \Delta_\gamma, +, 0, <)$, the Hahn product of the ordered groups Δ_γ.

We prove now the following theorem and explain later how it relates to the surjectivity of a logarithm.

Theorem 4.3 *Let Γ and Δ be totally ordered sets without greatest element, and fix an element $0 \in \Delta$. Suppose that Γ' is a cofinal subset of Γ and that $\iota\colon \Gamma' \to \Delta^\Gamma$ is an order preserving embedding. Then the image $\iota\Gamma'$ is not convex in Δ^Γ.*

Proof Since Δ has no greatest element, we can choose a map $\tau\colon \Delta \to \Delta$ such that $\tau\delta > \delta$ for all $\delta \in \Delta$. For every well ordered set $S \subset \Gamma$ and every

$d = (d_\gamma)_{\gamma \in \Gamma} \in \Delta^\Gamma$, set

$$d \oplus S = (d'_\gamma)_{\gamma \in \Gamma} \text{ where } d'_\gamma := \begin{cases} d_\gamma & \text{if } \gamma \notin S \\ \tau d_\gamma & \text{if } \gamma \in S \end{cases}.$$

Observe that the support of $d \oplus S$ is contained in support$(d) \cup S$ and thus, it is again well ordered. Further, if $S, S' \subset \Gamma$ are well ordered sets (or empty), then

$$S \subsetneq S' \Rightarrow d \oplus S < d \oplus S'. \tag{50}$$

Now let Γ' be a cofinal subset of Γ and suppose that $\iota \colon \Gamma' \to \Delta^\Gamma$ is an order preserving embedding such that the image $\iota\Gamma'$ is convex in Δ^Γ. We wish to deduce a contradiction. The idea of the proof is the following. Let ON denote the class of ordinal numbers. We shall define an infinite ON $\times$ $\mathbb{N}$ matrix (!) with coefficients in Γ', such that each column $(\gamma_\nu^{(n)})_{\nu \in \text{ON}}$ is a strictly increasing sequence in Γ'. Since Γ' is a *set*, every column of this matrix will provide a contradiction at the end of the construction (cf. figure).

$$\begin{pmatrix} \gamma_0^{(1)} & \cdots & \gamma_0^{(n)} & \gamma_0^{(n+1)} & \cdots \\ \vdots & & \vdots & \vdots & \\ \gamma_\nu^{(1)} & \cdots & \gamma_\nu^{(n)} & \gamma_\nu^{(n+1)} & \cdots \\ \vdots & & \vdots & \vdots & \\ \gamma_\mu^{(1)} & \cdots & \gamma_\mu^{(n)} = ? & \cdots & \cdots \\ \vdots & & & & \\ \cdots & \cdots & \cdots & \cdots & \cdots \end{pmatrix}$$

To get started, we have to define the first row of the matrix. That is, by induction on $n \in \mathbb{N}$, we define elements $\gamma_0^{(n)} \in \Gamma'$. For the definition of the succeeding rows, we shall use transfinite induction.

We choose an arbitrary $\gamma_0^{(1)} \in \Gamma'$. Having already constructed $\gamma_0^{(n)}$, we carry through the following induction step. Since Γ has no greatest element, the same holds for Γ', and there is some $\alpha^{(n)} \in \Gamma'$ such that $\gamma_0^{(n)} < \alpha^{(n)}$. Hence, $\iota\gamma_0^{(n)} < \iota\alpha^{(n)}$. Let $\beta^{(n)} \in \Gamma$ be the least element of support$(\iota\gamma_0^{(n)}) \cup$ support$(\iota\alpha^{(n)})$ for which

$$(\iota\gamma_0^{(n)})_{\beta^{(n)}} < (\iota\alpha^{(n)})_{\beta^{(n)}}.$$

Since Γ has no greatest element and Γ' is a cofinal subset, we can choose $\gamma_0^{(n+1)} \in \Gamma'$ such that $\beta^{(n)} < \gamma_0^{(n+1)}$.

If $S \subset \Gamma$ is a well ordered set with least element $\gamma_0^{(n+1)}$, then

$$\iota\gamma_0^{(n)} < \iota\gamma_0^{(n)} \oplus S < \iota\alpha^{(n)}. \tag{51}$$

Indeed, $(\iota\gamma_0^{(n)} \oplus S)_\beta = (\iota\gamma_0^{(n)})_\beta$ for every $\beta < \gamma_0^{(n+1)}$. In particular,

$$(\iota\gamma_0^{(n)} \oplus S)_{\beta^{(n)}} = (\iota\gamma_0^{(n)})_{\beta^{(n)}} < (\iota\alpha^{(n)})_{\beta^{(n)}},$$

which implies the second inequality of (51). Its first inequality follows from (50).

The image of Γ' in Δ^Γ being convex, (51) yields that also $\iota\gamma_0^{(n)} \oplus S$ lies in this image. Thus, $\iota^{-1}(\iota\gamma_0^{(n)} \oplus S)$ is a well defined element of Γ'.

2. A combinatorial result and its consequences

Suppose now that for some ordinal number $\mu \geq 1$ we have chosen elements $\gamma_\nu^{(n)} \in \Gamma'$, $\nu < \mu$, $n \in \mathbb{N}$, such that for every fixed n, the sequence $(\gamma_\nu^{(n)})_{\nu<\mu}$ is strictly increasing. Then we set

$$\gamma_\mu^{(n)} := \iota^{-1}(\iota\gamma_0^{(n)} \oplus \{\gamma_\nu^{(n+1)} \mid \nu < \mu\}) \in \Gamma'$$

for every $n \in \mathbb{N}$. If $\lambda < \mu$, then $\{\gamma_\nu^{(n+1)} \mid \nu < \lambda\} \subsetneq \{\gamma_\nu^{(n+1)} \mid \nu < \mu\}$ and thus, $\gamma_\lambda^{(n)} < \gamma_\mu^{(n)}$ by (50). So for every ordinal number μ, the sequences $(\gamma_\nu^{(n)})_{\nu<\mu}$ can be extended. We obtain strictly increasing sequences of arbitrary length, contradicting the fact that their length is bounded by the cardinality of Γ'. □

If in the hypothesis of this theorem we drop the condition that Γ has no greatest element, the situation changes drastically. Suitably chosen ordered sets Γ and Δ will even admit an isomorphism $\Gamma \simeq \Delta^\Gamma$. We study this situation and related questions in [K–K–S2].

Now we apply Theorem 4.3 to logarithmic cross-sections of the power series field $K = k((G))$ with canonical valuation $v_{\min}$ (cf. Chapter 1, Section 5). Note that $(K, v_{\min})$ has value group G and residue field k. Further, it is henselian. Consequently, $v_{\min}$ is convex with respect to every order $<$ on K (cf. [KN–WR]).

Theorem 4.4 *Let k be an ordered field and G a non-trivial ordered abelian group. Let $v_{\min}$ be the canonical valuation on $K = k((G))$ and $<$ any order on K. Then $(K,<)$ admits no surjective $v_{\min}$-logarithmic cross-section.*

Proof Recall from Section 5 of Chapter 1 that the canonical additive complement of the valuation ring $R_w = k[[G]]$ is the Hahn product $k((G^{<0})) = k^{G^{<0}}$. Since the complements are unique up to isomorphism, a surjective $v_{\min}$-logarithmic cross-section h would induce an isomorphism $G \simeq k((G^{<0}))$. This in turn would imply that $G^{<0}$ has no greatest element and would give rise to an embedding of $G^{<0}$ in $k^{G^{<0}}$ with convex image, which contradicts Theorem 4.3. □

This theorem implies Theorem 4.2. Indeed, a $v_{\min}$-compatible exponential of K would induce a surjective $v_{\min}$-logarithmic cross-section, which is impossible. If $(k,<)$ is archimedean, then $v_{\min}$ coincides with the natural valuation v of the ordered field $(k((G)),<)$. So the second assertion of Theorem 4.2 follows by Lemma 1.18.

Corollary 4.5 *Let k be an ordered field and G a non-trivial ordered abelian group. Let $<$ be any order on $K = k((G))$. Let w' be a coarsening of the canonical valuation $v_{\min}$. Then $(K,<)$ does not admit any w'-compatible exponential.*

Proof Since K is a power series field with canonical valuation $v_{\min}$, it can also be written as a power series field $(Kw')((w'(K)))$ with canonical valuation w'. From Theorem 4.2 it follows that no exponential can be compatible with w'. □

Recall that by [KA], we know that a valued field (K,w) of residue characteristic 0 is a power series field with canonical valuation w if and only if it is maximal. So we can restate our corollary as follows: *If the ordered field K admits an exponential f, then there is no non-trivial coarsening w' of its natural valuation v which is maximally valued and compatible with f.* We prove the following generalization:

Theorem 4.6 *Let f be an exponential on the ordered field K and w a coarsening of the natural valuation v of K such that f is compatible with w. Then there is* no coarsening $\tilde{w}$ of w such that the valuation $\overline{w} = w/\tilde{w}$ induced by w on the residue field $K\tilde{w}$ is non-trivial and $(K\tilde{w}, \overline{w})$ is maximally valued.

Proof Suppose to the contrary that such a coarsening $\tilde{w}$ exists. We have that $R_w \subset R_{\tilde{w}}$. Let $\overline{\mathbf{A}}$ be a group complement of R_w in $R_{\tilde{w}}$ and $\tilde{\mathbf{A}}$ a group complement of $R_{\tilde{w}}$ in $(K, +, 0, <)$. Then $\tilde{\mathbf{A}} \amalg \overline{\mathbf{A}}$ is a group complement of R_w in $(K, +, 0, <)$. Further, f induces an isomorphism h from $G = w(K)$ onto $\tilde{\mathbf{A}} \amalg \overline{\mathbf{A}}$ as ordered groups. In particular, $G^{<0}$ has no greatest element.

The value group of $\overline{w}$ is isomorphic to a non-trivial convex subgroup $\overline{G}$ of G. Since $(K\tilde{w}, \overline{w})$ is maximally valued and has residue field $(K\tilde{w})(w/\tilde{w}) = Kw$, it is isomorphic to the power series field $(Kw)((\overline{G}))$. Hence, $\overline{\mathbf{A}}$ is isomorphic to a Hahn product $(Kw)^{\overline{G}^{<0}}$. This yields an embedding of the non-trivial convex subgroup $H := \overline{G} \cap h^{-1}(\overline{\mathbf{A}})$ of $\overline{G}$ in $(Kw)^{\overline{G}^{<0}}$. Under this embedding, the image of the final segment $H^{<0}$ of $\overline{G}^{<0}$ is convex in $(Kw)^{\overline{G}^{<0}}$. But $\overline{G}^{<0}$ is a final segment of $G^{<0}$ and thus has no greatest element. This contradicts Theorem 4.3. □

Remark 4.7 (i) An argument similar to that used in establishing Theorem 4.4 shows that a non-trivial Hahn group (i.e. a maximally valued group) cannot be an exponential group, hence cannot be the natural value group of an exponential field. Thus Hahn groups never admit natural contractions.

(ii) Hyper-real fields admit exponentials (cf. [G–J]). Our result shows that hyper-real fields are never maximally valued.

(iii) In [MI] it is shown that an o-minimal expansion of the reals is either polynomially bounded (cf. Chapter 6), or else the exponential is definable. Combined with our result, we deduce that an o-minimal expansion of the reals which has a power series model, is necessarily polynomially bounded.

3 Existence of logarithmic cross-sections

We want to establish our positive result (Corollary 4.11). The crucial tool here are group cross-sections, which we introduce now. Let G be an ordered abelian group. A **group cross-section** is an embedding $s : v_G(G) \to G^{<0}$ of ordered sets such that $v_G \circ s$ is the identity on $v_G(G)$.

Remark 4.8 Note that every ordered abelian group G admits a group cross-section: For $\gamma \in v_G(G)$, we just have to set $s(\gamma) = g$ where $g \in G^{<0}$ is an arbitrary element of value $v_G(g) = \gamma$ (s is order preserving by (7)).

For example, if Γ is any totally ordered set and R any ordered Abelian group, then the map s_0 defined by $\Gamma \ni \gamma \mapsto -1_\gamma \in R^\Gamma$ is a group cross-section for $G = R^\Gamma$ (cf. Section 5 of Chapter 1). We call s_0 the **canonical group cross-section** of R^Γ.

We first establish the following easy lemma:

Lemma 4.9 *Let R be any ordered Abelian group and Γ, Γ' be totally ordered sets. Then any embedding ϑ of Γ in Γ' lifts canonically to an embedding $\hat{\vartheta}$ of the Hahn group R^Γ in the Hahn group $R^{\Gamma'}$. Moreover, $\hat{\vartheta}(R^\Gamma)$ is the subgroup $\mathbb{R}^{\vartheta(\Gamma)}$ consisting of all sequences whose support lies completely in $\vartheta(\Gamma)$.*

Proof Set $H = R^\Gamma$ and $H' = R^{\Gamma'}$. Let $g \in H$. Define $\hat{\vartheta}(g) \in R^{\Gamma'}$ as follows. For $\gamma' \in \Gamma'$, set

$$(\hat{\vartheta}(g))(\gamma') = g(\gamma)$$

3. Existence of logarithmic cross-sections

if $\gamma' = \vartheta(\gamma)$ for some $\gamma \in \Gamma$, and

$$(\hat{\vartheta}(g))(\gamma') = 0$$

if $\gamma' \notin \operatorname{im}(\vartheta)$. An easy computation establishes that $v_{H'}(\hat{\vartheta}(g)) = \vartheta(v_H(g))$, that is, $\hat{\vartheta}$ lifts ϑ. The second assertion of the lemma is clear. □

Below let G be an ordered abelian group, denote its value set by Γ and its archimedean components by B_γ (recall that B_γ embeds in $\mathbb{R}$). Recall that Hahn's Embedding Theorem states that there is an order preserving group embedding ρ of G in the Hahn product $\mathbf{H}_{\gamma \in \Gamma} B_\gamma$, provided that G is divisible (cf. Theorem 0.26).

Proposition 4.10 *Let G be a non-trivial divisible ordered abelian group, k an ordered field. Assume that every archimedean component of G embeds in the ordered additive group of k. Then any group cross-section of G lifts to a $v_{\min}$-logarithmic cross-section of $k((G))$.*

In particular, every group cross-section of G lifts to a v-logarithmic cross-section of $\mathbb{R}((G))$.

Proof Let $s : \Gamma \to G^{<0}$ be a group cross-section. By Lemma 4.9, s lifts to an embedding $\hat{s} : k^\Gamma \to k^{G^{<0}}$. If every archimedean component B_γ of G embeds in $(k, +, 0, <)$, then there is an embedding $\tau : \mathbf{H}_{\gamma \in \Gamma} B_\gamma \to \mathbf{H}_\Gamma(k, +, 0, <) = k^\Gamma$. So $h = \hat{s} \circ \tau \circ \rho$ is the required $v_{\min}$-logarithmic cross-section for $k((G))$. Since every archimedean ordered abelian group embeds in $\mathbb{R}$, the second assertion follows from the first. □

Corollary 4.11 *If G is non-trivial and divisible, then the real closed field $\mathbb{R}((G))$ admits a v-compatible prelogarithm.*

Proof In Section 5 of Chapter 1, we have already shown that a v-right logarithm on $\mathbb{R}((G))$ always exists. Also the residue field admits a logarithm, whereas the v-logarithmic cross-section gives rise to a v-left prelogarithm. The assertion now follows by Theorem 4.1. □

Before we give an example, we need to recall the following well-known fact which follows from Neumann's Lemma (cf. [FU], [N]).

Lemma 4.12 *Let $1 + I_v$ be the group of 1-units of the valuation ring $\mathbb{R}[[G]]$. Then for every $\varepsilon \in I_v$,*

$$\ell_R(1+\varepsilon) = \sum_{i=1}^{\infty} (-1)^{(i-1)} \frac{\varepsilon^i}{i} \tag{52}$$

is a canonically defined element of I_v and ℓ_R is a surjective v-right logarithm. "Canonical" means in particular: if $G \subset G'$ and ℓ'_R is defined on the 1-units of $\mathbb{R}[[G']]$ in the same way, then it extends ℓ_R.

Example 4.13 Let $\Gamma = \mathbb{Z}$ and $G = \mathbb{R}^\mathbb{Z}$. Then by the above corollary, the power series field

$$K_0 = \mathbb{R}((\mathbb{R}^\mathbb{Z}))$$

carries a basic v-compatible prelogarithm, which we denote by $\log_0$. We now want to compute an explicit formula for $\log_0(a)$, for positive $a \in K_0$.

Let s_0 be the canonical group cross-section of $\mathbb{R}^{\mathbb{Z}}$, so $s_0(z) = -1_z$ for every $z \in \mathbb{Z}$. For $g \in \mathbb{R}^{\mathbb{Z}}$, we have that $r_z := g(z) \in \mathbb{R}$ for every $z \in \mathbb{Z}$. Instead of viewing g as a map, let us work with the more suggestive expression

$$g = \sum_{z \in \mathbb{Z}} r_z 1_z.$$

Although this sum may be infinite, it has a canonical interpretation in the Hahn product $\mathbb{R}^{\mathbb{Z}}$.

As in Section 5 of Chapter 1, when we work in K_0, we will write t^g instead of 1_g. Hence, every element of K_0 can be written in the form

$$\sum_{g \in G} r_g t^g$$

with $r_g \in \mathbb{R}$ and well-ordered support $\{g \in G \mid r_g \neq 0\}$. Recall that the natural valuation on K_0 is given by $vt^g = g$. Recall that $\mathrm{Mon}\,(K_0)$ is the canonical multiplicative complement to the group of positive units, and that $\mathrm{Mon}\,(K_0)$ is isomorphic to G through the isomorphism $(-v)(t^g) = -g$. Recall also that $\mathrm{Neg}\,(K_0)$ is the canonical additive complement (cf. Section 5 of Chapter 1). We will construct a v-left prelogarithm mapping $\mathrm{Mon}\,(K_0)$ into $\mathrm{Neg}\,(K_0)$.

If $a \in K_0$ is positive, we can write

$$a = t^g r(1 + \varepsilon)$$

with $r \in \mathbb{R}^{>0}$, $g \in G$, and ε an infinitesimal. Then for any prelogarithm $\log_0$ we must have that

$$\log_0 a = \log_0 t^g + \log_0 r + \log_0 (1 + \varepsilon).$$

We define $\log_0 r = \log r$, where $\log$ is the natural logarithm on $\mathbb{R}^{>0}$. We define the v-right logarithm $\log_0(1+\varepsilon)$ through the Taylor expansion of $\log(1+x)$ (cf. Lemma 4.12 above):

$$\log_0(1+\varepsilon) = \sum_{i=1}^{\infty} (-1)^{(i-1)} \frac{\varepsilon^i}{i}. \tag{53}$$

It remains to give an appropriate definition for $\log_0 t^g$, using the method described above in this section. Let h_0 be the v-logarithmic cross-section obtained by lifting the canonical group cross-section s_0, that is, $h_0 = \hat{s}_0$ (cf. Lemma 4.9). For $g \in G$, write $g = \sum_{z \in \mathbb{Z}} r_z 1_z$. We compute:

$$h_0(-g) = h_0\left(-\left(\sum_{z \in \mathbb{Z}} r_z 1_z\right)\right) = \sum_{z \in \mathbb{Z}} -r_z t^{-1_z}.$$

Now on $\mathrm{Mon}\,(K_0)$,

$$\log_0 = h_0 \circ (-v).$$

Thus,

$$\log_0(t^g) = h_0(-g) = \sum_{z \in \mathbb{Z}} -r_z t^{-1_z}.$$

Obviously, $\log_0(t^g) \in \mathrm{Neg}\,(K_0)$. Our formula now reads

$$\log_0 a = \sum_{z \in \mathbb{Z}} -r_z t^{-1_z} + \log r + \sum_{i=1}^{\infty} (-1)^{(i-1)} \frac{\varepsilon^i}{i}.$$

Note that $\log_0$ is obviously not surjective: if $g \in G^{<0}$ is not of the form $g = -1_z$, then t^g is not in the image of $\log_0$. In other words, for $g \in G^{<0}$, t^g is in the image of $\log_0$ if and only if g is in the image of the embedding of $\mathbb{Z}$ in $G^{<0}$ given by $z \mapsto -1_z$. Hence, we have to enlarge $\mathbb{Z}$ to get every t^g into the image. We describe how to do this in the next section.

Note also that the prelogarithm $\log_0$ thus constructed on the power series field $\mathbb{R}((\mathbb{R}^{\mathbb{Z}}))$ *certainly does not* satisfy the growth axiom! Indeed, if we compute $\zeta_{\log_0}$ (cf. Section 4 of Chapter 3) we find that it is the identity map. Thus by (43), $\log_0$ is not a (GA)-prelogarithm.

In the next chapter, we describe a canonical method to modify $\log_0$ in order to obtain a (GA)-prelogarithm on the field K_0.

4 From prelogarithms to logarithms

Using Corollary 4.11, we shall now construct non-archimedean exponential fields which are countable unions of power series fields. Indeed, a common method to obtain surjectivity of a map is to construct the union over a suitable countably infinite chain of fields.

- **Construction of a surjective v-logarithmic cross-section.**

To get started, let Γ_0 be any non-empty totally ordered set and $G_0 = \mathbb{R}^{\Gamma_0}$ the Hahn group. Set $K_0 = \mathbb{R}((G_0))$. Let s_0 be the canonical group cross-section of G_0 and $\text{Neg}(K_0)$ be the canonical group complement of $\mathbb{R}[[G_0]]$ in K_0 (see figure). Let

$$h_0 : G_0 \to \text{Neg}(K_0)$$

be the v-logarithmic cross-section $\hat{s}_0$ of K_0 obtained by lifting s_0. Now assume that we have already constructed G_{n-1}, K_{n-1}, and the v-logarithmic cross-section

$$h_{n-1} : G_{n-1} \to \text{Neg}(K_{n-1}).$$

Since G_{n-1} is isomorphic to a subgroup of $\text{Neg}(K_{n-1})$ through h_{n-1}, we can take G_n to be a group containing G_{n-1} as a subgroup and admitting an isomorphism h_n onto $\text{Neg}(K_{n-1})$ which extends h_{n-1}. We set

$$K_n := \mathbb{R}((G_n)).$$

Hence, $K_{n-1} \subset K_n$ canonically (the elements of K_{n-1} being those elements of K_n whose support is a subset of G_{n-1}). Further, the canonical complement $\text{Neg}(K_n)$ for the valuation ring $\mathbb{R}[[G_n]]$ contains $\text{Neg}(K_{n-1})$. Thus, h_n is an embedding of G_n in $\text{Neg}(K_n)$ which extends h_{n-1} and satisfies

$$h_n(G_n) = \text{Neg}(K_{n-1}). \tag{54}$$

74   4. Construction of exponential fields

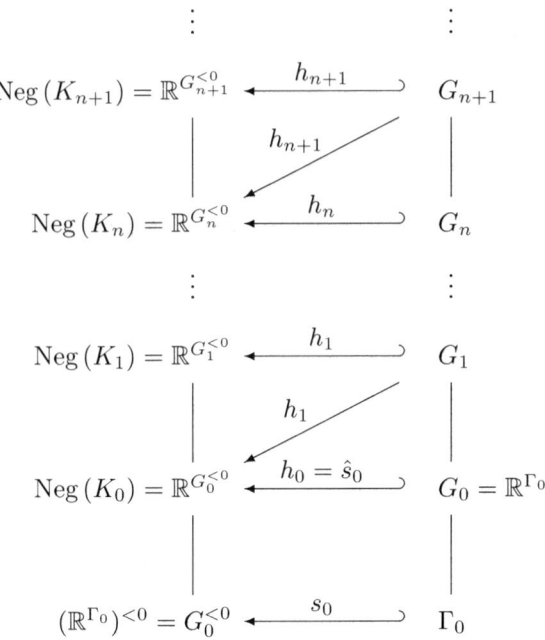

By induction on n, we obtain a chain of power series fields K_n, $n \in \mathbb{N}$. Now set
$$K_\omega := \bigcup_{n \in \mathbb{N}} K_n \quad \text{and} \quad h_\omega := \bigcup_{n \in \mathbb{N}} h_n \ .$$
Also the groups G_n form a chain, and their union $G_\omega := \bigcup_{n \in \mathbb{N}} G_n$ is the value group of K_ω. We have that $K_\omega \subset \mathbb{R}((G_\omega))$. Similarly, the group complements $\text{Neg}(K_n)$ form a chain, and their union $\text{Neg}(K_\omega) := \bigcup_{n \in \mathbb{N}} \text{Neg}(K_n)$ is a group complement of the valuation ring $\bigcup_{n \in \mathbb{N}} \mathbb{R}[[G_n]] = \mathbb{R}[[G_\omega]] \cap K_\omega$ in K_ω.

Finally, by (54), $h_\omega : G_\omega \to \text{Neg}(K_\omega)$ is surjective. The surjective v-logarithmic cross-section h_ω gives us the v-left logarithm $h_\omega \circ (-v)$ on K_ω.

• **Construction of a logarithm on K_ω.**

For every $n \in \mathbb{N}$, we now define the logarithm ℓ_n on the positive units of K_n just by the formula (53) above:
$$\ell_n(u) = \ell(r(1+\varepsilon)) = \log(r) + \sum_{i=1}^{\infty} (-1)^{(i-1)} \frac{\varepsilon^i}{i} \ .$$
Then ℓ_{n+1} is an extension of ℓ_n to the positive units of K_{n+1}. Hence, $\ell_\omega := \bigcup_{n \in \mathbb{N}} \ell_n$ maps the positive units of K_ω onto the valuation ring of K_ω. Now for positive $a = t^g r(1+\varepsilon) \in K_\omega$, our logarithm is given by
$$\log_0 a = h_\omega(-g) + \log r + \sum_{i=1}^{\infty} (-1)^{(i-1)} \frac{\varepsilon^i}{i} \ .$$
This completes our construction.

Note that $\text{Mon}(K_\omega) = \cup \text{Mon}(K_n)$ and $\text{Neg}(K_\omega) = \cup \text{Neg}(K_n)$. Therefore, $\log_0(\text{Mon}(K_\omega)) = \text{Neg}(K_\omega)$. Note also that $\log_0$ extends the prelogarithm that

4. From prelogarithms to logarithms

we started with on K_0. By the observation at the end of the example above, we deduce that $\log_0$ is *not* a (GA)-logarithm. We will improve $\log_0$ in the next chapter to obtain a (GA)-logarithm that will moreover satisfy all the axioms of real exponentiation.

We denote the non-archimedean real closed field thus constructed by $\mathbb{R}((\Gamma_0))^{EL}$, and call it the exponential-logarithmic power series field over Γ_0. We shall see in the next chapter that this field is extremely rich. It admits a multitude of exponentials of distinct growth rates.

CHAPTER 5

Models for the elementary theory of the reals with restricted analytic functions and exponentiation

In this chapter, we show how to endow our exponential-logarithmic power series field $\mathbb{R}((\Gamma_0))^{EL}$ with a (GA)-logarithm. This is done by twisting the canonical group cross-section of R^{Γ_0} through an increasing automorphism of Γ_0.

We shall give several explicit examples, exhibiting the connection between endomorphisms of the exponential rank and the growth rate of the constructed exponentials. This allows us to construct on a fixed real closed field infinitely many exponentials of distinct exponential ranks. Thus, in contrast to the rank, the exponential rank of a real closed exponential field is in general not uniquely determined.

1 Twisting a group cross-section by an automorphism

We have seen in the last section that the prelogarithm $\log_0$ does not satisfy (GA). We will deal with this problem now.

We need the following notion. Let K be a non-archimedean ordered field, and set $v(K) = G$ and $v_G(G) = \Gamma$. A v-logarithmic cross-section h on K is a **strong v logarithmic cross-section** if

$$v(h(g)) > g \quad \text{for every } g \in G^{<0}.$$

Recall that every embedding (resp. isomorphism) of ordered abelian groups induces canonically an embedding (resp. isomorphism) of their value sets as ordered sets. In particular, a v-logarithmic cross-section h induces an embedding $\tilde{h}$ such that the following diagram commutes:

$$\begin{array}{ccc} G & \xrightarrow{h} & A \\ \downarrow{v_G} & & \downarrow{v} \\ v_G(G) & \xrightarrow{\tilde{h}} & G^{<0} \end{array}$$

If h is onto, then so is $\tilde{h}$ (in that case, $\tilde{h}$ is a group exponential, cf. Section 2 of Chapter 2.) We have that

$$\tilde{h}(v_G(g)) > g \Leftrightarrow v(h(g)) > g$$

for every $g \in G^{<0}$.

We see that h is a strong v-logarithmic cross-section if and only if

$$\tilde{h}(v_G(g)) > g \qquad \text{for all } g \in G^{<0}. \tag{55}$$

We can now encode (GA) in the logarithmic cross-section:

Lemma 5.1 *Let ℓ be a v-compatible prelogarithm on K. Then ℓ is a (GA)-prelogarithm if and only if h_ℓ is a strong v-logarithmic cross-section.*

Proof Let $\tilde{h}_\ell : \Gamma \to G^{<0}$ be the embedding induced by h_ℓ. An easy computation establishes that:

$$\chi_\ell = \tilde{h}_\ell \circ v_G$$

and

$$\zeta_\ell = v_G \circ \tilde{h}_\ell.$$

Now the assertion of the lemma follows immediately from Lemma 3.19. □

Lemma 5.1 does not yet tell anything about the existence of (strong) logarithmic cross-sections. We will now discuss this problem. Let σ be an automorphism of a totally ordered set S. We say that σ is an **increasing automorphism** if $\sigma(x) > x$ for all $x \in S$.

Proposition 5.2 (Twisting a group cross-section) *Let G be an ordered Abelian group, and σ an increasing automorphism of $v_G(G)$. Then given any group cross-section s of G, the embedding $\tilde{h} := (s \circ \sigma) : v_G(G) \to G^{<0}$ satisfies that $\tilde{h}(v_G(g)) > g$ for all $g \in G^{<0}$.*

Proof Indeed, $v_G(\tilde{h}(v_G(g))) = \sigma(v_G(g)) > v_G(g)$ and thus $\tilde{h}(v_G(g)) > g$ for $g \in G^{<0}$. □

Note that there are plenty of groups whose value set admits increasing automorphisms. For instance, this is the case if $v_G(G)$ is isomorphic to an arbitrary non-zero ordered abelian group, as an ordered set.

Now the question arises whether an embedding $\tilde{h}$ can be lifted to a v-logarithmic cross-section h. We observe below that a lifting always exists if K is a power series field. Let Γ be any totally ordered set, and set $G = \mathbb{R}^\Gamma$. Let $K = \mathbb{R}((G))$ be the power series field with exponents in G. By Lemma 4.9 and the results of the last chapter, we see that *every* embedding of ordered sets $\tilde{h} : \Gamma \to G^{<0}$ lifts to a v-logarithmic cross-section $h : G = \mathbb{R}^\Gamma \to \text{Neg}(K)$. In particular, every embedding of ordered sets $\tilde{h} : \Gamma \to G^{<0}$ which satisfies (55) lifts to a strong v-logarithmic cross-section. We can now get the desired strong version of Corollary 4.11:

Corollary 5.3 *Let G be a non-trivial divisible ordered Abelian group. Let Γ be its value set. Then every increasing automorphism of Γ induces a (GA)-v-compatible prelogarithm on the real closed field $K = \mathbb{R}((G))$.*

Proof Let σ be the automorphism. Choose any group cross-section s of G. By Lemma 5.2 the embedding $\tilde{h} := s \circ \sigma : v_G(G) \to G^{<0}$ satisfies that $\tilde{h}(v_G(g)) > g$ for all $g \in G^{<0}$. By the observation preceding the corollary, $\tilde{h}$ lifts to a strong v-logarithmic cross-section h. We define the v-left prelogarithm by $h \circ (-v)$. We take the v-middle logarithm to be the real log on $\mathbb{R}$, and the v-right logarithm as in Lemma 4.12. By Theorem 4.1, we obtain a v-compatible prelogarithm ℓ on K. By Lemma 5.1, ℓ is a (GA)-prelogarithm. □

In the next section, we will exploit Corollary 5.3 to construct exponential-logarithmic power series fields which are models of $T_{\text{an}}(\exp)$, the theory of the reals with restricted analytic functions and exponentiation (cf. [D–M–M1] for an axiomatization of this theory and Chapter 6 for more details).

2 The exponential-logarithmic power series field

In this section, we shall construct canonically the exponential-logarithmic power series field $\mathbb{R}((\Gamma_0))^{EL(\sigma)}$, a model of $T_{\text{an}}(\exp)$ (see chapter 6, sections 7 and 8 for the definitions and basic properties of the theories T_{an} and $T_{\text{an}}(\exp)$).

To get started, let Γ_0 be any non-empty totally ordered set which admits increasing automorphisms. Fix such an automorphism σ. Set $G_0 := \mathbb{R}^{\Gamma_0}$ and $K_0 = \mathbb{R}((G_0))$. Let s_0 be the canonical group cross-section of G_0 and $\text{Neg}(K_0)$ be the canonical group complement of $\mathbb{R}[[G_0]]$ in K_0. Let

$$h_0^\sigma : G_0 \to \text{Neg}(K_0)$$

be the strong v-logarithmic cross-section of K_0 obtained by lifting

$$\tilde{h} := s_0 \circ \sigma$$

(cf. Proposition 5.2 and Corollary 5.3).

Exactly as we have done in Section 4 of Chapter 4, we construct a surjective v-logarithmic cross-section h_ω^σ, but this time, we make sure that h_ω^σ is a *strong* v-logarithmic cross-section, in order to get a (GA)-logarithm at the end of the procedure.

So assume that we have already constructed G_{n-1}^σ, K_{n-1}^σ, and the strong v-logarithmic cross-section

$$h_{n-1}^\sigma : G_{n-1}^\sigma \to \text{Neg}(K_{n-1}^\sigma)$$

satisfying

$$v(h_{n-1}^\sigma(g)) > g \quad \text{for all } g \in (G_{n-1}^\sigma)^{<0}. \tag{56}$$

Again, since G_{n-1}^σ is isomorphic to a subgroup of $\text{Neg}(K_{n-1}^\sigma)$ through h_{n-1}^σ, we can take G_n^σ to be a group containing G_{n-1}^σ as a subgroup and admitting an isomorphism h_n^σ onto $\text{Neg}(K_{n-1}^\sigma)$ which extends h_{n-1}^σ. We set

$$K_n^\sigma := \mathbb{R}((G_n^\sigma)).$$

As before, $K_{n-1}^\sigma \subset K_n^\sigma$ canonically. So h_n^σ is an embedding of G_n^σ into $\text{Neg}(K_n^\sigma)$ which extends h_{n-1}^σ.

We show that h_n^σ is again a strong logarithmic cross-section. For $g \in G_n^\sigma$, the image $h_n^\sigma(g)$ lies in $\text{Neg}(K_{n-1}^\sigma)$, and $v(h_n^\sigma(g))$ lies in its value set $(G_{n-1}^\sigma)^{<0}$. Consequently, in (56) we may replace $g \in (G_{n-1}^\sigma)^{<0}$ by $v(h_n^\sigma(g))$ for $g \in (G_n^\sigma)^{<0}$. But $v(h_{n-1}^\sigma(v(h_n^\sigma(g)))) > v(h_n^\sigma(g))$ implies that $h_{n-1}^\sigma(v(h_n^\sigma(g))) > h_n^\sigma(g)$, because $h_n^\sigma(g) < 0$ and $h_{n-1}^\sigma(v(h_n^\sigma(g))) < 0$. Since h_n^σ extends h_{n-1}^σ, this may be read as $h_n^\sigma(v(h_n^\sigma(g))) > h_n^\sigma(g)$. Since h_n^σ is order preserving, this implies $v(h_n^\sigma(g)) > g$. Thus, we have proved that (56) holds with n in the place of $n-1$.

By our induction on n, we obtain a chain of fields K_n^σ, $n \in \mathbb{N}$. As before, we take

$$K_\omega^\sigma := \bigcup_{n \in \mathbb{N}} K_n^\sigma$$

and

$$h_\omega^\sigma := \bigcup_{n \in \mathbb{N}} h_n^\sigma.$$

So

$$h_\omega^\sigma : G_\omega^\sigma \to \text{Neg}(K_\omega^\sigma)$$

is surjective. Moreover, h_ω^σ is a strong v-logarithmic cross-section since (56) holds for all n.

To define the v-right logarithm and complete our construction, we now proceed exactly as in Section 4 of Chapter 4: For every $n \in \mathbb{N}$, we define ℓ_n mapping the positive units of K_n^σ onto the valuation ring of K_n^σ by

$$\ell_n(u) = \ell(r(1+\varepsilon)) = \log(r) + \sum_{i=1}^\infty (-1)^{(i-1)} \frac{\varepsilon^i}{i}$$

(where log is the logarithm on $\mathbb{R}$). Then ℓ_{n+1} is an extension of ℓ_n to the positive units of K_{n+1}^σ. Hence,

$$\ell_\omega := \bigcup_{n \in \mathbb{N}} \ell_n$$

is well-defined and ℓ_ω maps the positive units of K_ω^σ onto the valuation ring of K_ω^σ.

For positive $a = t^g r(1+\varepsilon) \in K_\omega^\sigma$, our logarithm is given by

$$\log_\sigma(a) = h_\omega^\sigma(-g) + \log r + \sum_{i=1}^\infty (-1)^{(n-1)} \frac{\varepsilon^i}{i} \ .$$

Now $\log_\sigma$ is a (GA)-v-compatible logarithm on K_ω^σ by Lemma 5.1.

But K_ω^σ carries an even much richer structure, as we explain now.

- **Model theoretic properties of K_ω^σ.**

We shall now show that K_ω^σ can be naturally made into a model of $T_{\mathrm{an}}(\exp)$. Below, let $T(\exp)$ be the elementary theory of the reals with the real exponential function exp, and $T(\exp|[-1,1])$ be the elementary theory of the reals with the real exponential restricted to the unit box (cf. Chapter 6 for details).

A real closed ordered field K with exponential f is a model of $T(\exp)$ if it is a model of $T(\exp|[-1,1])$ and f satisfies (GA). This is the content of Ressayre's Theorem (cf. [RE]). Ressayre's Theorem also holds if one adds the restricted analytic functions (cf. [D–M–M1], (4.10)). So let us note:

Lemma 5.4 *Let K be a model of the reals with restricted analytic functions, and f an exponential on K such that f coincides on $[-1,1]$ with the interpretation of the (symbol for the) restricted exponential. Then (K, f) is a model of $T_{\mathrm{an}}(\exp)$ if and only if $\ell = f^{-1}$ is a (GA)-logarithm.*

In [D–M–M1], a canonical way is described of how to make any power series field $\mathbb{R}((G))$ into a model of the theory T_{an} of the reals with restricted analytic functions, if G is divisible. This is done by defining the restricted analytic functions via their Taylor expansions using Neumann's Lemma (exactly as we have done it for $\log_\sigma$ on the group of positive units).

This definition is canonical in the same spirit as in Lemma 4.12. Hence it is compatible with the inclusions $K_n^\sigma \subset K_{n+1}^\sigma$. Moreover, it makes every K_n^σ into a model of T_{an}. By the model completeness of T_{an} (cf. [D]), $K_n^\sigma \prec K_{n+1}^\sigma$ (that is, K_n^σ is an elementary submodel of K_{n+1}^σ) for every n. Hence, K_ω^σ is the union over an elementary chain of models K_n^σ of T_{an} and is thus itself a model of T_{an} (cf. [C–K]).

Recall that $\ell_\omega = \bigcup_{n \in \mathbb{N}} \ell_n$. We have noted above that every ℓ_n is an isomorphism from the positive units of K_n^σ onto the valuation ring $\mathbb{R}[[G_n^\sigma]]$. This is just

2. The exponential-logarithmic power series field

because ℓ_n has a compositional inverse defined on $\mathbb{R}[[G_n^\sigma]]$ by:

$$f_n(r+\varepsilon) = \exp(r) \cdot \sum_{i=0}^{\infty} \frac{\varepsilon^i}{i!}.$$

where exp is the exponential function on $\mathbb{R}$ (cf. [N]). Set $f_\sigma = \log_\sigma^{-1}$. Clearly f_n it is the restriction of f_σ to $\mathbb{R}[[G_n^\sigma]]$. Let $n \in \mathbb{N}$ and a be an element of the interval $[-1,1]$ of K_n^σ. Since $[-1,1] \subset \mathbb{R}[[G_n^\sigma]]$, we can write $a = r + \varepsilon$ with $r \in \mathbb{R}$ and $v(\varepsilon) > 0$, and we have:

$$f_\sigma(a) = f_n(a) = \exp(r) \cdot \sum_{i=0}^{\infty} \frac{\varepsilon^i}{i!}.$$

Therefore, f_σ coincides on the $[-1,1]$ of K_n^σ with the interpretation of the restricted exp (given by its Taylor expansion), for every n. Hence, this is also true on the interval $[-1,1]$ of K_ω^σ. From Lemma 5.4 we conclude that $(K_\omega^\sigma, f_\sigma)$ is a model of $T_{\mathrm{an}}(\exp)$. Note that as before we have: $\log_\sigma(\mathrm{Mon}\,(K_\omega^\sigma)) = \mathrm{Neg}\,(K_\omega^\sigma)$.

We denote the model of $T_{\mathrm{an}}(\exp)$ thus constructed by $\mathbb{R}((\Gamma_0))^{EL(\sigma)}$, and call it the exponential-logarithmic power series field over (Γ_0, σ).

Remark 5.5 (Extension to the algebraic closure) In view of the work in [Z1] and [Z2], it is interesting to construct algebraically closed fields with an exponential function. Consider the exponential f_σ on the field $K_\omega^\sigma \subset \mathbb{C}((G_\omega^\sigma))$ which was constructed above. We wish to extend f_σ to the algebraic closure $\widetilde{K_\omega^\sigma}$ of K_ω^σ. Since K_ω^σ is real closed, we have that $\widetilde{K_\omega^\sigma} = K_\omega^\sigma[i]$, where $i = +\sqrt{-1}$. Since we want to obtain the equality $f_\sigma(a+ib) = f_\sigma(a) \cdot f_\sigma(ib)$ for all $a, b \in K_\omega^\sigma$, we only have to define $f_\sigma(ib)$.

We consider $\mathrm{Neg}\,(K_\omega^\sigma)$. We set $f_\sigma(ib) := 1$ if $b \in \mathrm{Neg}\,(K_\omega^\sigma)$. It remains to define $f_\sigma(ib)$ for b in the valuation ring of K_ω^σ. Every such b can be written as $r + \varepsilon$ with $r \in \mathbb{R}$ and $v(\varepsilon) > 0$. Let exp denote the exponential function on $\mathbb{C}$. In view of $f_\sigma(ib) = f_\sigma(ir) \cdot f_\sigma(i\varepsilon)$ with $f_\sigma(ir) = \exp(ir) \in \mathbb{C} \subset \widetilde{K_\omega^\sigma}$, it remains to define $f_\sigma(i\varepsilon)$. We define sin and cos on the infinitesimals ε by their Taylor expansion. Since $\sin^2(x) + \cos^2(x) = 1$ is an identity of convergent power series, we have that also $\sin^2(\varepsilon) + \cos^2(\varepsilon) = 1$ for every infinitesimal ε. Now we set $f_\sigma(i\varepsilon) := \cos(\varepsilon) + i\sin(\varepsilon)$.

We are now going to establish a surprising result about exponential-logarithmic power series fields.

Theorem 5.6 *Let Γ_0 be any totally ordered set and σ any increasing automorphism of Γ_0. Then*

$$\mathbb{R}((\Gamma_0))^{EL(\sigma)} \simeq \mathbb{R}((\Gamma_0))^{EL}$$

as ordered fields.

Proof Clearly, it is enough to show that for all $n \in \mathbb{N}$ we have isomorphisms of ordered groups

$$i_n : G_n \to G_n^\sigma,$$

such that i_n extends i_{n-1}. It will then immediately follow that

$$G_\omega \simeq G_\omega^\sigma$$

through the isomorphism $i_\omega = \cup i_n$ and thus,

$$K_\omega \simeq K_\omega^\sigma.$$

82   5. Models for the reals with restricted analytic functions and exponentiation

For n=0 there is nothing to show since $G_0 = G_0^\sigma$ by construction. Let us also observe that since σ is an automorphism, we have $s_0(\Gamma_0) = (s_0 \circ \sigma)(\Gamma_0)$. It follows that
$$h_0(G_0) = h_0^\sigma(G_0) .$$
Indeed, by Lemma 4.9 we compute:
$$\begin{aligned} h_0(G_0) &= h_0(\mathbb{R}^{\Gamma_0}) = \hat{s}_0(\mathbb{R}^{\Gamma_0}) = \mathbb{R}^{s_0(\Gamma_0)} = \mathbb{R}^{(s_0 \circ \sigma)(\Gamma_0)} \\ &= \widehat{(s_0 \circ \sigma)}(\mathbb{R}^{\Gamma_0}) = h_0^\sigma(\mathbb{R}^{\Gamma_0}) = h_0^\sigma(G_0) , \end{aligned}$$
as required.

Set $i_0 = (h_0^\sigma)^{-1} \circ h_0$. Now assume by induction on $n \in \mathbb{N}$ that
$$i_n = (h_n^\sigma)^{-1} \circ \hat{i}_{n-1} \circ h_n : G_n \to G_n^\sigma$$
is an isomorphism of ordered groups such that
$$i_n | G_{n-1} = i_{n-1}$$
and
$$\widehat{i_{n-1}} : \mathrm{Neg}\,(K_{n-1}) \to \mathrm{Neg}\,(K_{n-1}^\sigma)$$
is the lifting of the isomorphism
$$i_{n-1} : (G_{n-1})^{<0} \to (G_{n-1}^\sigma)^{<0}$$
(cf. Lemma 4.9). Recall that by construction we have that
$$h_{n+1}(G_{n+1}) = \mathrm{Neg}\,(K_n)$$
and similarly,
$$h_{n+1}^\sigma(G_{n+1}^\sigma) = \mathrm{Neg}\,(K_n^\sigma) .$$
We set
$$i_{n+1} = (h_{n+1}^\sigma)^{-1} \circ \hat{i}_n \circ h_{n+1} ,$$
where
$$\hat{i}_n : \mathrm{Neg}\,(K_n) \to \mathrm{Neg}\,(K_n^\sigma)$$
is the lifting of the isomorphism
$$i_n : (G_n)^{<0} \to (G_n^\sigma)^{<0} .$$
Obviously, i_{n+1} being the compositum of three isomorphisms of ordered groups is itself an isomorphism of ordered groups. We must now verify that $i_{n+1}|G_n = i_n$. Note that by construction we have that $h_{n+1}|G_n = h_n$ and $h_{n+1}^\sigma|G_n^\sigma = h_n^\sigma$. This implies that for all $y \in h_n^\sigma(G_n^\sigma)$,
$$(h_{n+1}^\sigma)^{-1}(y) = (h_n^\sigma)^{-1}(y) .$$
For $g \in G_n$, we compute:
$$i_{n+1}(g) = (h_{n+1}^\sigma)^{-1} \circ \hat{i}_n \circ h_{n+1}(g) = ((h_{n+1}^\sigma)^{-1} \circ \hat{i}_n)(h_n(g)) .$$
Now by induction hypothesis $i_n|G_{n-1} = i_{n-1}$, thus clearly $\hat{i}_n|\mathrm{Neg}\,(K_{n-1}) = \widehat{i_{n-1}}$. On the other hand, $h_n(g) \in \mathrm{Neg}\,(K_{n-1})$. Thus,
$$\hat{i}_n(h_n(g)) = \widehat{i_{n-1}}(h_n(g)).$$
Now $\widehat{i_{n-1}}(h_n(g)) \in \mathrm{Neg}\,(K_{n-1}^\sigma) = h_n^\sigma(G_n^\sigma)$. Thus,
$$(h_{n+1}^\sigma)^{-1}(\widehat{i_{n-1}}(h_n(g))) = (h_n^\sigma)^{-1}(\widehat{i_{n-1}}(h_n(g))) = i_n(g).$$

This completes the proof. □

A major consequence of this theorem is that we can now always work with our canonically defined exponential-logarithmic power series field $\mathbb{R}((\Gamma_0))^{EL}$. The following corollary illustrates the strength of the theorem. It gives the result that we promised at the end of the last chapter.

Theorem 5.7 *Let Γ_0 be any non-empty totally ordered set. Then the real closed field $\mathbb{R}((\Gamma_0))^{EL}$ admits*

(i) the logarithm $\log_0$ for which the exponential field $(\mathbb{R}((\Gamma_0))^{EL}, (\log_0)^{-1})$ is not a model of $T(\exp)$.

(ii) for every increasing automorphism σ of Γ_0 a canonically defined logarithm $\log_\sigma$ for which the exponential field $(\mathbb{R}((\Gamma_0))^{EL}, (\log_\sigma)^{-1})$ is a model of $T_{\mathrm{an}}(\exp)$.

Thus, $\mathbb{R}((\Gamma_0))^{EL}$ admits at least as many distinct logarithms $\log_\sigma$ for which $(\mathbb{R}((\Gamma_0))^{EL}, (\log_\sigma)^{-1})$ is a model of $T_{\mathrm{an}}(\exp)$, as there are distinct increasing automorphisms of Γ_0. Furthermore, the exponential $(\log_0)^{-1}$ agrees with the exponential $(\log_\sigma)^{-1}$ on the valuation ring of $\mathbb{R}((\Gamma_0))^{EL}$, for every increasing automorphism σ of Γ_0.

Proof We have already seen in the last chapter that $(\mathbb{R}((\Gamma_0))^{EL}, (\log_0)^{-1})$ is not a model of $T(\exp)$. We have seen above that $(\mathbb{R}((\Gamma_0))^{EL(\sigma)}, (\log_\sigma^{\cdot})^{-1}))$ is a model of $T_{\mathrm{an}}(\exp)$. Now the assertion follows by Theorem 5.6: we just copy the canonically defined logarithm $\log_\sigma$ from $\mathbb{R}((\Gamma_0))^{EL(\sigma)}$ over to $\mathbb{R}((\Gamma_0))^{EL}$, using the isomorphism i_ω. This canonically defines a logarithm on $\mathbb{R}((\Gamma_0))^{EL}$ which we shall continue to denote by $\log_\sigma$. □

In the next section, we will get yet a stronger result. Indeed, $\mathbb{R}((\Gamma_0))^{EL}$ admits not only many distinct exponentials, but moreover many exponentials of *distinct exponential ranks*.

3 Models of arbitrary principal exponential rank

Remark 5.8 Power series fields provide a straightforward method of constructing non-archimedean ordered real closed fields of any given principal rank τ, provided that we can construct a divisible ordered abelian group of principal rank τ. But the latter is easy: we just take the Hahn group $G = \mathbb{R}^{\tau^*}$ (cf. Section 1 of Chapter 3 and recall that τ^* is τ endowed with the reversed ordering). The principal rank of the real closed field $K = \mathbb{R}((G))$ will then be τ.

For the construction of exponential fields with arbitrary given principal exponential rank, this approach fails since power series fields never admit exponentials (cf. Chapter 4). Nevertheless, we will construct exponential fields with arbitrary principal exponential rank. Instead of using power series fields, we will use the exponential-logarithmic power series fields.

Let Γ be a totally ordered set and σ an increasing automorphism of Γ. We define the σ-**rank** of (Γ, σ) to be the order type of the chain $\Gamma/\sim_\sigma$ (cf. Remark 3.20). We start by the following improvement of Corollary 5.3:

Corollary 5.9 *Let G be a non-trivial divisible ordered Abelian group. Let Γ be its value set. Let σ be any increasing automorphism of Γ. Denote by τ the σ-rank of (Γ, σ). Then σ induces a (GA)-v-compatible prelogarithm ℓ on the real closed field $K = \mathbb{R}((G))$. Furthermore, the principal exponential rank of $(\mathbb{R}((G)), \ell)$ is precisely τ^*.*

Proof Let σ be the automorphism. Choose any group cross-section s of G. Set $\tilde{h} := s \circ \sigma$, and lift $\tilde{h}$ to a strong v-logarithmic cross-section h. Define the v-left prelogarithm by $h \circ (-v)$. (As in the proof of Corollary 5.3 we take the v-middle logarithm to be the real log on $\mathbb{R}$, and the v-right logarithm as in Lemma 4.12). We obtain a (GA)-v-compatible prelogarithm ℓ on K. We compute:
$$\zeta_\ell = v_G \circ \tilde{h} = v_G \circ s \circ \sigma = (v_G \circ s) \circ \sigma = \sigma$$
(since s is a group cross-section). The assertion now follows by Corollary 3.35. □

Example 5.10 We reconsider the field
$$K_0 = \mathbb{R}((\mathbb{R}^{\mathbb{Z}}))$$
that we studied in the example at the end of Section 3. Consider the increasing automorphism of $\mathbb{Z}$ given by $\sigma(z) = z + 1$. Obviously the σ-rank of $(\mathbb{Z}, \sigma)$ is 1. By the above corollary, the power series field admits a (GA)-v-compatible prelogarithm $\log_\sigma$ of principal exponential rank 1. In particular, of exponential rank 1. We now want to compute an explicit formula for $\log_\sigma(a)$, for positive $a \in K_0$. Let s_0 be the canonical group cross-section of $\mathbb{R}^{\mathbb{Z}}$, so $s_0(z) = -1_z$ for every $z \in \mathbb{Z}$. Thus $\tilde{h}_0(z) = (s_0 \circ \sigma)(z) = -1_{z+1}$ for every $z \in \mathbb{Z}$. Let h_0 be the (GA)-v-logarithmic cross-section obtained by lifting $s_0 \circ \sigma$, that is, $h_0 = \widehat{s_0 \circ \sigma}$. Let $a \in K_0$ be positive, and write
$$a = t^g r(1 + \varepsilon)$$
with $r \in \mathbb{R}^{>0}$, $g \in G$, and ε an infinitesimal. For $g \in G$, write $g = \sum_{z \in \mathbb{Z}} r_z 1_z$. We compute:
$$h_0(-g) = h_0\left(-\left(\sum_{z \in \mathbb{Z}} r_z 1_z\right)\right) = \sum_{z \in \mathbb{Z}} -r_z t^{-1_{z+1}}.$$
Now on $\text{Mon}(K_0)$ we have that
$$\log_\sigma = h_0 \circ (-v).$$
Thus,
$$\log_\sigma(t^g) = h_0(-g) = \sum_{z \in \mathbb{Z}} -r_z t^{-1_{z+1}}.$$
Our formula now reads:
$$\log_\sigma(a) = \sum_{z \in \mathbb{Z}} -r_z t^{-1_{z+1}} + \log r + \sum_{i=1}^{\infty} (-1)^{(i-1)} \frac{\varepsilon^i}{i}.$$

Obviously again, $\log_\sigma$ is not yet surjective, so we show the following theorem. It is the promised strengthening of Theorem 5.7.

Theorem 5.11 *Let Γ_0 be any non-empty totally ordered set. Let σ be any increasing automorphism of Γ_0. Set $\tau_\sigma =$ the σ-rank of (Γ_0, σ). Then σ induces canonically a (GA)-v-compatible logarithm $\log_\sigma$ on the real closed field $\mathbb{R}((\Gamma_0))^{EL}$, for which the exponential field $(\mathbb{R}((\Gamma_0))^{EL}, (\log_\sigma)^{-1})$ is a model of $T_{\text{an}}(\exp)$. Furthermore, the principal exponential rank of $(\mathbb{R}((\Gamma_0))^{EL}, (\log_\sigma)^{-1})$ is precisely $(\tau_\sigma)^*$.*

Thus, $\mathbb{R}((\Gamma_0))^{EL}$ admits at least as many logarithms $\log_\sigma$ of pairwise distinct exponential ranks (for which $(\mathbb{R}((\Gamma_0))^{EL}, (\log_\sigma)^{-1})$ is a model of $T_{\text{an}}(\exp)$), as there are increasing automorphisms σ of Γ_0 of pairwise distinct σ-ranks. Furthermore, the exponentials $\log_\sigma^{-1}$ all agree on the valuation ring of $\mathbb{R}((\Gamma_0))^{EL}$.

3. Models of arbitrary principal exponential rank

Proof Let σ be given. By Theorem 5.6, we can work with $\mathbb{R}((\Gamma_0))^{EL(\sigma)}$ with its canonically defined logarithm $\log_\sigma$. Let ℓ_0^σ be the prelogarithm induced by h_0^σ on K_0^σ. By Corollary 5.9, $(K_0^\sigma, \ell_0^\sigma)$ has principal exponential rank equal to τ_σ. Note that by construction, $\log_\sigma |K_0^\sigma = \ell_0^\sigma$. We wish to show that $(K_\omega^\sigma, \log_\sigma)$ has the same principal exponential rank as $(K_0^\sigma, \ell_0^\sigma)$ (that is, the extension preserves the principal exponential rank).

Let $a \in K_\omega^\sigma$ be positive infinite. We claim that a is $\log_\sigma$-equivalent to some $b_n \in K_0^\sigma$. Write $a = b.u$ with u a positive unit, and b a monomial. Thus $v(a) = v(b)$, so a is archimedean, and hence a fortiori $\log_\sigma$-equivalent to b. Thus it is enough to establish our claim for $b \in \mathrm{Mon}\,(K_\omega^\sigma) = \cup \mathrm{Mon}\,(K_n^\sigma)$. Now there is $n \in \mathbb{N}$ such that $b \in \mathrm{Mon}\,(K_n^\sigma)$. Set $b = b_0$. By construction, the image of h_n is $\mathrm{Neg}\,(K_{n-1}^\sigma)$. Consequently, $\log_\sigma b_0 \in \mathrm{Neg}\,(K_{n-1}^\sigma)$. Now $\log_\sigma b_0$ is positive infinite, thus again there is a $b_1 \in \mathrm{Mon}\,(K_{n-1}^\sigma)$ such that $\log_\sigma b_0$ is archimedean, and hence a fortiori $\log_\sigma$-equivalent to b_1. Clearly b_0 is $\log_\sigma$-equivalent to $\log_\sigma b_0$. By transitivity, b_0 is $\log_\sigma$-equivalent to b_1. By induction on n, we find that b_0 is $\log_\sigma$-equivalent to some infinite positive element $b_n \in \mathrm{Mon}\,(K_0^\sigma)$. This proves our assertion. □

Curiously enough, we shall use exactly the same type of argument in the next chapter, to establish that the Hardy field associated to a polynomially bounded + (exp) expansion of the reals has levels (provided that the restricted exp is definable in the polynomially bounded expansion).

We now have the exponential analogue to the situation described in the remark at the beginning of this section.

Theorem 5.12 *Let τ be any order type. Then there exists a totally ordered set Γ_0 which admits an increasing automorphism σ of σ-rank equal to τ^*. Consequently, the power series field $\mathbb{R}((\mathbb{R}^{\Gamma_0}))$ admits a (GA)-v-compatible prelogarithm of principal exponential rank τ. Moreover, the exponential-logarithmic power series field $(\mathbb{R}((\Gamma_0))^{EL}, (\log_\sigma)^{-1})$ has principal exponential rank τ.*

Proof Let T be an ordered set having order type τ^*. We may assume that τ is non-trivial, that is, $T \neq \emptyset$, since otherwise, we could set $G = \{0\}$ and $\mathbb{R}((G)) = \mathbb{R}$, and the usual logarithm would do the job. We define the ordered set Γ_0 to be the sum (in the sense of ordered sets) of copies of $\mathbb{Z}$ over the index set T. That is, $\Gamma_0 = T \amalg \mathbb{Z}$ lexicographically ordered. We let σ be the automorphism which sends an element n in any of these copies to its successor $n+1$ in the same copy. Clearly, the σ-rank of (Γ_0, σ) is equal to τ^*. The two other assertions follow from Corollary 5.9 and Theorem 5.11, together with the observation that $(\tau^*)^* = \tau$. □

Example 5.13 Let $\Gamma_0 = \mathbb{R}^\mathbb{N}$. Then Γ_0 has $\aleph_0$ many automorphisms of distinct σ-rank. Indeed, define σ_n as follows:

$$\sigma_n(\gamma)(m) = \begin{cases} \gamma(m) & \text{if } m \neq n \\ \gamma(m) + 1 & \text{if } m = n. \end{cases} \qquad (57)$$

That is, σ_n is obtained by translating the n-th coefficient by 1 and leaving all other fixed. Note that σ_n is an increasing automorphism of $\mathbb{R}^\mathbb{N}$. We see that σ_1 has σ-rank equal 1, whereas σ_2 has σ-rank $\mathbb{R}$. By induction, σ_n has σ-rank $\mathbb{R}^{(n-1)}$, lexicographically ordered. Now observe that $\mathbb{R}^n$ is not isomorphic to $\mathbb{R}^m$ if $m \neq n$ (cf. [KS2]).

Thus the σ_n's have pairwise distinct σ-ranks. It follows that $\{\log_{\sigma_n} \mid n \in \mathbb{N}\}$ is a set of $\aleph_0$ many logarithms of pairwise distinct exponential ranks, on the exponential-logarithmic power series field $\mathbb{R}((\mathbb{R}^\mathbb{N}))^{EL}$. Further, $(\mathbb{R}((\mathbb{R}^\mathbb{N}))^{EL}, (\log_{\sigma_n})^{-1})$ is a model of $T_{\mathrm{an}}(\exp)$, for every $n \in \mathbb{N}$.

Remark 5.14 Actually, we do not need to start with a Γ_0 which admits many increasing automorphisms in order to get logarithms of distinct exponential ranks. Increasing automorphisms are already produced through the construction procedure of the exponential-logarithmic power series field, over any non-empty totally ordered set. We illustrate below.

Even if we start with $\Gamma_0 = \{1\}$, we can obtain countable infinitely many exponentials with distinct exponential ranks, which turn the exponential-logarithmic power series field

$$\mathbb{R}((\{1\}))^{EL}$$

into a model of real exponentiation.

Indeed, by a slight modification of our above construction, one can let $\Gamma_n = v_G(G_n)$ play the role of Γ_0. More precisely, having already constructed the field $\mathbb{R}((\{1\}))^{EL}$ as above using the canonical group cross-section, we now set up to define new logarithms on it. This is done as follows. We start working with h_n instead of h_0 to construct the logarithm. Then, we already have the index set Γ_n at hand. We claim that Γ_n admits at least n increasing order automorphisms of distinct principal exponential ranks. This is seen as follows. On $G_0 = \mathbb{R}$ we take the order automorphism σ_1 to be the multiplication by $1/2$. It is an increasing order automorphism on $\Gamma_1 = \mathbb{R}^{<0}$ and lifts to an order automorphism on $G_1 \simeq \mathbb{R}^{\Gamma_1}$. But on G_1 there is also the order automorphism σ_2 induced through multiplication by $1/2$ on the components $\mathbb{R}$. The restrictions of σ_1 and σ_2 to $G_1^{<0} = \Gamma_2$ are increasing automorphisms of Γ_2 of distinct σ-ranks. Indeed, while for every $\gamma \in \Gamma_2$, the sequence $\sigma_1^k \gamma$, $k \in \mathbb{N}$, is cofinal in Γ_2, the sequence $\sigma_2^k \gamma$, $k \in \mathbb{N}$, is not. Indeed, σ_1 has σ-rank 1, whilst σ_2 has σ-rank Γ_1. Just observe that for $\gamma_1, \gamma_2 \in \Gamma_1$ the elements -1_{γ_1} and -1_{γ_2} of Γ_2 are not σ_2-equivalent. On the other hand, if $x \in \Gamma_2$ and $\min \mathrm{support}(x) = \gamma$ (for $\gamma \in \Gamma_1$), then x is σ_2-equivalent to -1_γ. Thus $\{-1_\gamma \mid \gamma \in \Gamma_1\}$ forms a complete system of representatives of the σ_2-classes of Γ_2, and this set is obviously order isomorphic to Γ_1. We now have obviously proven that σ_1 and σ_2 have distinct σ-ranks.

We now consider Γ_3. By the same argument as above, σ_1 and σ_2 lift to increasing automorphisms $\hat{\sigma}_1$ and $\hat{\sigma}_2$ of Γ_3 of σ-ranks 1 and Γ_1 respectively. But on G_2 there is also the order automorphism σ_3 induced through multiplication by $1/2$ on the components $\mathbb{R}$, and again the restriction of σ_3 to $G_2^{<0} = \Gamma_3$ is an increasing automorphism of Γ_3 of σ-rank equal to Γ_2. Now by Theorem 4.3, Γ_1 is not isomorphic to Γ_2. Thus $\sigma_1, \sigma_2, \sigma_3$ have pairwise distinct σ-ranks.

By induction, we obtain n increasing order automorphisms on Γ_n, of σ-ranks one, $\Gamma_1, \ldots, \Gamma_{n-1}$ respectively. By Theorem 4.3, Γ_m is not isomorphic to Γ_n if $n \neq m$, and we have established our claim.

Letting $n \in \mathbb{N}$ vary and using the n essentially different increasing order automorphisms of Γ_n for every n, we obtain constructions of $\aleph_0$ many exponentials with distinct exponential ranks.

Remark 5.15 The theory of real closed fields and $T_{\mathrm{an}}(\exp)$ are both o-minimal (cf. [P-S] and [D-M-M1]). Nevertheless, the order type of the models is a too weak

3. Models of arbitrary principal exponential rank

invariant. Indeed, in [A–K] it was shown that for every uncountable cardinal κ there are 2^κ real closed fields of cardinality κ, all isomorphic as ordered groups, but not as ordered fields. The above examples provide an exponential analogue of this result, where again, the exponential function is considered instead of the squaring function. This answers a question of Zil'ber concerning o-minimal structures to the negative.

We close this chapter by the following theorem, which may be viewed as an exponential analogue of Kaplansky's Embedding Theorem for real closed fields. A **truncation closed embedding** is an embedding such that the truncation of any power series in the image of the embedding lies again in this image.

Theorem 5.16 *Let (K, f) be a model of $T_{\mathrm{an}}(\exp)$ with principal exponential rank τ. Then (K, f) can be elementarily embedded in a model (K_ω, f_ω) of $T_{\mathrm{an}}(\exp)$ which is a countable union of power series fields and has principal exponential rank τ. The embedding can be chosen to be truncation closed.*

Proof Set $v(K) = G_0$. By [D–M–M1], $\mathbb{R}((G_0))$ is a model of the theory T_{an} of the reals with restricted analytic functions. Moreover, there is a truncation closed embedding of K in $\mathbb{R}((G_0))$ which respects the restricted analytic functions. Now the v-left logarithm of K induces canonically a strong v-logarithmic cross-section h_0 on $K_0 = \mathbb{R}((G_0))$. We continue the construction exactly as we did at the beginning of this section. The so-obtained exponential f_ω on K_ω extends f. By the model completeness of T_{an} (cf. [D]) and the results of [D–M–M1], the embedding of (K, f) in (K_ω, f_ω) is elementary. □

CHAPTER 6

Exponential Hardy fields

In the past years, the study of Hardy fields has become very interesting and fruitful. M. Rosenlicht studied Hardy fields of various finite ranks (cf. [RO2]).

Here we consider Hardy fields which contain the germ of the real exponential function and are closed under composition. These are non-archimedean exponential fields, and as such, their ranks are very large and not easy to classify. The exponential rank is then a suitable tool for the classification of such exponential Hardy fields. For example, Hardy fields arising from "polynomially bounded + (exp)" expansions of the reals have exponential rank 1 (cf. Section 4).

Complementing Rosenlicht's work on Hardy fields of finite rank, in the next sections we will compute the principal ranks of some important exponential Hardy fields. Moreover, we shall describe their value groups and residue fields, and derive structure theorems for those Hardy fields from our results.

We will apply our knowledge of the residue fields to give a solution to a problem by Hardy in asymptotic analysis and to show that the restriction of the Riemann ζ-function to $(1, \infty)$ is not definable in the expansion of the reals by restricted analytic functions, exp and log (cf. Section 10). Both results were first obtained by different methods in [D–M–M2].

For the model theoretic notions appearing in this chapter, we refer the reader to [C–K]. For details on o-minimality and o-minimal structures, cf. [D1] and [P–S]. In the next section, we gather some basic valuation theoretic facts that we shall need.

1 Some basic valuation theory

Throughout this chapter, v will always denote the natural valuation, unless stated otherwise.

Remark 6.1 Let L be a real closed field and w a convex valuation on L. Then the residue field Lw is a real closed field, and it can be embedded in L in such a way that the composition of the residue map with the embedding yields the identity on Lw. Indeed, the embedding can be constructed by use of Hensel's Lemma (a convex valuation on a real closed field is always henselian), since the residue field has characteristic 0 (cf. [PR1]). The image under every such embedding is a maximal subfield of R_w, and conversely, every maximal subfield K of R_w is isomorphic to Lw via the residue map. Therefore, we will always write $Lw = K \subset R_w$, where the equality is to be understood modulo the isomorphism induced by the residue map.

Lemma 6.2 *The maximal subfields of R_w are characterized by the property that for all $y \in R_w$ there is a unique $z \in K$ such that $w(y - z) > 0$. This implies the following: assume that $(F, w) \subset (L, w)$ is an extension of valued fields, with* embedded residue fields Fw, Lw such that $Fw \subset Lw$. If $z \in Lw$, $y \in F$ and $w(z - y) > 0$, then necessarily $z \in Fw$.

Take any ordered field L with convex valuation w. We denote the real closure of L by L^r. Then every embedding of Lw in L has a unique extension to an embedding of $(Lw)^r$ in L^r. (cf. [PR1]).

Lemma 6.3 *Let w be any valuation on $K(x_i \mid i \in I_1 \cup I_2)$ such that the values $w(x_i)$, $i \in I_1$, are rationally independent over $w(K)$, and the residues $x_i w$, $i \in I_2$, are algebraically independent over Kw. Then the elements x_i, $i \in I_1 \cup I_2$ are algebraically independent over K. Moreover,*

$$w(K(x_i \mid i \in I_1 \cup I_2)) = w(K) \oplus \bigoplus_{i \in I_1} \mathbb{Z} w(x_i)$$

$$K(x_i \mid i \in I_1 \cup I_2)w = Kw(x_i w \mid i \in I_2).$$

For the proof, see [KF3] or [BOU], Chapter VI, Section 10.3, Theorem 1.

Corollary 6.4 *Suppose that $\mathbb{R}(x_i \mid i \in I)$ is an ordered field such that the values $v(x_i)$, $i \in I$ are rationally independent. Let w be a convex valuation on $\mathbb{R}(x_i \mid i \in I)$. Assume that there is a subset $I_w \subset I$ such that $w(x_i) = 0$ for all $i \in I_w$ and that the values $w(x_i)$, $i \in I \setminus I_w$ are rationally independent. Then*

$$w(\mathbb{R}(x_i \mid i \in I)) = \bigoplus_{i \in I \setminus I_w} \mathbb{Z} w(x_i) \quad \text{and} \quad \mathbb{R}(x_i \mid i \in I)w = \mathbb{R}(x_i \mid i \in I_w).$$

Proof Let H_w denote the convex subgroup associated to w. For $i \in I_w$, $w(x_i) = 0$ implies that $v(x_i) \in H_w$. By the foregoing lemma, $v(\mathbb{R}(x_i \mid i \in I_w)) = \bigoplus_{i \in I_w} \mathbb{Z} v(x_i) \subset H_w$.

This proves that w is trivial on $\mathbb{R}(x_i \mid i \in I_w)$. So we can assume that the residue map is the identity on $\mathbb{R}(x_i \mid i \in I_w)$. Now apply the foregoing lemma with $K = \mathbb{R}(x_i \mid i \in I_w)$ (then $Kw = K$), $I_1 = I \setminus I_w$ and $I_2 = \emptyset$. $\square$

Recall from Chapter 0 that a sequence of elements $a_\nu \in K$, $\nu < \lambda$ (λ some limit ordinal), is pseudo Cauchy in (K, w) if $w(a_\rho - a_\sigma) < w(a_\sigma - a_\tau)$ for all ρ, σ, τ with $\rho < \sigma < \tau < \lambda$. It follows from the ultrametric inequality that $w(a_\nu - a_\tau) = w(a_\nu - a_{\nu+1})$ whenever $\nu < \tau < \lambda$. The element a is a (pseudo) limit of this pseudo Cauchy sequence if $w(a_\nu - a) = w(a_\nu - a_{\nu+1})$ for all $\nu < \lambda$. In general, there may be several distinct limits:

Lemma 6.5 *Let a be a limit of $(a_\nu)_{\nu < \lambda}$. Then b is also a limit of $(a_\nu)_{\nu < \lambda}$ if and only if $w(a - b) > w(a_\nu - a_{\nu+1})$ for all $\nu < \lambda$.*

Recall from Chapter 1 that an extension $(K, w) \subset (L, w)$ of valued fields is immediate if the canonical embedding of $w(K)$ in $w(L)$ and the canonical embedding of Kw in Lw are surjective (we then write $w(K) = w(L)$ and $Kw = Lw$). The henselization of a valued field is an immediate extension.

Lemma 6.6 *Assume that $(K, w) \subset (L, w)$ is immediate and that $a \in L \setminus K$. Then there is a pseudo Cauchy sequence in (K, w) with limit a, but not having a limit in K.*

The next lemma follows from the Lemma of Ostrowski (cf. [RI], [KF3]) and the results of Kaplansky's important paper [KA]:

1. Some basic valuation theory

Lemma 6.7 *Let K be any real closed field and w a convex valuation on K. Assume that $(a_\nu)_{\nu<\lambda}$ is a pseudo Cauchy sequence in (K,w), not having a limit in K. Assume further that in some extension of (K,w), there exists a limit a. Then the extension of w to $K(a)$ is uniquely determined and immediate.*

If $(K_1, w) \subset (K_2, w)$ is an immediate algebraic extension of ordered fields with convex valuation w, then their henselizations (in a fixed henselian extension field) are equal.

If the values $w(a_\nu - a_{\nu+1})$ are cofinal in $w(K)$, then $(a_\nu)_{\nu<\lambda}$ is called a **Cauchy sequence** in (K, w). Lemma 6.5 shows that if this sequence has a limit in K, then this limit is uniquely determined. Indeed, if $a, b \in K$ are limits, then $w(a - b) > w(K)$, that is, $w(a - b) = \infty$, or in other words, $a = b$. All elements in the completion of a valued field are limits of Cauchy sequences (and in particular, the completion is an immediate extension). Conversely:

Lemma 6.8 *Let the situation be as in Lemma 6.7, with $(a_\nu)_{\nu<\lambda}$ a Cauchy sequence. Then there is a unique embedding of $(K(a), w)$ over K in the completion of (K, w).*

Note that if $w(K)$ is archimedean, then it follows from Newton's method together with this lemma that the henselization of (K, w) is embeddable in the completion of (K, w). If w and v are arbitrary valuations such that v is finer than w and $Kw \subset K$, then (K, v) is henselian if and only if (K, w) and (Kw, v) are henselian (cf. [RI] or [KF3]). From these facts, one obtains:

Lemma 6.9 *Let K be an ordered field with convex valuation w. Suppose that $Kw \subset K$ and that (Kw, v) is henselian. Then the henselization of K with respect to v is equal to the henselization of K with respect to w. If in addition $w(K)$ is archimedean, this henselization is embeddable in the completion of (K, w).*

For the proof of the next lemma, see [PR1] or [KF3].

Lemma 6.10 *Let K be an ordered field with convex valuation w. Then K is real closed if and only if (K, w) is henselian, $w(K)$ is divisible and Kw is real closed. Further, $w(K^{\mathrm{r}}) = \mathbb{Q} \otimes w(K)$ (the divisible hull of $w(K)$), and $K^{\mathrm{r}}w = (Kw)^{\mathrm{r}}$. If $w(K)$ is divisible and Kw is real closed, then the real closure of K is equal to the henselization of K with respect to w (and embeddable in the completion of (K, w) if $w(K)$ is archimedean).*

Assume that $(M, \exp)$ is a model of $T(\exp)$, the elementary theory of the reals with exponentiation. Below we once more summarize some valuation theoretic properties that the exponential $\exp$ of M will satisfy. These properties will be deduced from the fact that $\exp$ of M is necessarily a (GAT)-exponential.

The exponential $\exp$ of M is an order preserving isomorphism from the additive group of M onto its multiplicative group of positive elements. Its inverse is the logarithm $\log$; it is order preserving and defined for all positive elements. Now assume that $(M, \exp)$ is an elementary extension of $(\mathbb{R}, \exp)$. Consequently, if $z \in M$ is positive infinite, that is, $z > \mathbb{R}$, then $\log(z) > \log(\{r \in \mathbb{R} \mid r > 0\}) = \mathbb{R}$. In other words,

$$v(z) < 0 \land z > 0 \Rightarrow v(\log(z)) < 0 \land \log(z) > 0. \tag{58}$$

Further, $\exp$ satisfies the Taylor axiom scheme (T). The following lemma follows from the results of Chapter 1 and Chapter 3:

Lemma 6.11 *Assume that $(M, \exp)$ is an elementary extension of $(\mathbb{R}, \exp)$, and let w be a convex valuation on M. Then*
(i) $\exp$ is a v-compatible exponential on M.
(ii) $\exp(I_w) = 1 + I_w$.
(iii) $\log(\mathcal{U}_w^{>0}) \subset R_w$.
(iv) $v_G(v(\log(a))) > v_G(v(a))$ for all $a \in \mathbf{P}_M$.

Proof Since $(M, \exp)$ is an elementary extension we have: $v(\exp(1) - 1) = v(e - 1) = 0$. Thus $\exp$ is v-compatible by Lemma 1.17. Since $\exp$ is v-compatible and T_1, (ii) follows by Lemma 3.17. Since $\exp$ is also (GA), (iii) follows from Lemma 3.13. Since $\exp$ is (GA), the induced contraction is centripetal (cf. Lemma 3.19) and (iv) follows by Lemma A.8. $\square$

From this lemma, we obtain the following equations which we shall refer to: for every $z \in M$,

$$w(z) > 0 \Rightarrow w(\exp(z)) = 0 \wedge w(\exp(z) - 1) = w(z) \qquad (59)$$
$$v(z) = 0 \Rightarrow v(\exp(z)) = 0 \qquad (60)$$
$$w(z) = 0 \wedge z > 0 \Rightarrow w(\log(z)) \geq 0 \qquad (61)$$
$$v(z) \geq 0 \Leftrightarrow v(\exp(z)) = 0. \qquad (62)$$

and for all $m \in \mathbb{N}$,

$$v(z) < 0 \wedge z > 0 \Rightarrow v(\exp(z)) < mv(z) < v(z) < mv(\log(z)) < 0. \qquad (63)$$

Recall from Chapter 1 that the valuation v is a homomorphism from the multiplicative group $M^{>0}$ of positive elements onto the value group $v(M)$. Its kernel is $\mathcal{U}_v^{>0} = \{z \in M \mid v(z) = 0 \wedge z > 0\}$. So v induces an isomorphism $M^{>0}/\mathcal{U}_v^{>0} \simeq v(M)$, which is order reversing by convexity of the valuation. The exponential $\exp$ is an order preserving isomorphism from the additive group of M onto the multiplicative group $M^{>0}$. By (62), the preimage of $\mathcal{U}_v^{>0}$ under $\exp$ is precisely R_v. Hence,

Lemma 6.12 *The map $z \mapsto v(\exp(-z))$ induces an order preserving isomorphism $M/R_v \simeq v(M)$ of ordered abelian groups. In particular, if $v(a) < 0$, then the map $\mathbb{R} \ni r \mapsto v(\exp(-ra)) \in v(M)$ is order preserving.*

If the elements z_j, $j \in J$, are rationally independent over R_v in the additive group of M, then the values $v(\exp(z_j))$, $j \in J$, are rationally independent in $v(M)$.

2 Hardy fields

We start with a review of notions and terminology from Rosenlicht's papers [RO1], [RO2] and [RO3].

Let f, g be real valued functions defined on positive halflines of $\mathbb{R}$ (that is, their domain is either $\mathbb{R}$, or an interval of the form $[a, \infty)$ or (a, ∞) for some $a \in \mathbb{R}$). We say that f and g have the same **germ at** ∞ if $f(x) = g(x)$ **ultimately**: that is, if there exists $N \in \mathbb{N}$ such that $x \geq N$ implies $f(x) = g(x)$. Throughout this chapter, we shall identify f with its germ.

A **Hardy field** H is a set of germs at ∞ of real valued functions defined on positive halflines of $\mathbb{R}$ which is closed under differentiation and forms a field under pointwise addition and multiplication of germs.

Let $f \neq 0$, $f \in H$. Since $1/f \in H$, $f(x) \neq 0$ ultimately. Since the derivative $f' \in H$, f is ultimately differentiable, and thus ultimately continuous. It follows

2. Hardy fields

that f is ultimately always positive or always negative. This key property of Hardy fields allows us to regard them as ordered fields in the following way: If $f, g \in H$ we define $f > g$ if $f - g$ is ultimately positive.

Example 6.13 The fields $\mathbb{Q}$, $\mathbb{R}$ are Hardy fields consisting just of constant germs. They are archimedean. Let x denote the germ of the identity function. Then obviously $x > \mathbb{R}$ (that is, $x > r$, for every constant germ $r \in \mathbb{R}$). The Hardy field $\mathbb{R}(x)$ is thus non-archimedean.

In the sequel, we will be interested only in Hardy fields containing $\mathbb{R}(x)$.

If $f \in H$, then by the argument above we see that, if non-zero, f' is ultimately strictly positive or strictly negative. Thus f is ultimately monotonic. Thus $\lim_{x \to \infty} f(x)$ exists as an element of $\mathbb{R} \cup \{+\infty, -\infty\}$. Hence a Hardy field never contains oscillating functions like $\sin(x)$. This observation will allow us to introduce a valuation on H. Let $f, g \in H$ non-zero. Set $f \simeq g$ if and only if

$$\lim_{x \to \infty} f(x)/g(x)$$

is a *non-zero* real number. This is an equivalence relation. Denote the equivalence class of f by $v(f)$. That is,

$$v(f) = v(g) \Leftrightarrow \lim_{x \to \infty} f(x)/g(x) \in \mathbb{R} \setminus \{0\} .$$

Define an addition on the set of classes $v(H)$: $v(f) + v(g) = v(fg)$. Define an ordering on $v(H)$ by

$$v(f) > v(g) \Leftrightarrow \lim_{x \to \infty} f(x)/g(x) = 0 .$$

Then the map

$$f \mapsto v(f)$$

is a convex valuation on H. Rosenlicht calls v the canonical valuation on H. The proof of the following lemma is straightforward:

Lemma 6.14 *The canonical valuation on a Hardy field is (equivalent to) the natural valuation.*

It is interesting to describe the valuation ring of finite elements and the ideal of infinitesimals in this case in terms of limits. It is easily seen that

$$R_v = \{f \mid \lim_{x \to \infty} f(x) \in \mathbb{R}\}$$

and

$$I_v = \{f \mid \lim_{x \to \infty} f(x) = 0\} .$$

Obviously, the units of R_v are given by

$$\mathcal{U}_v = \{f \mid \lim_{x \to \infty} f(x) \in \mathbb{R} \setminus \{0\}\},$$

whereas the set of positive infinite elements of H is $\mathbf{P}_H = \{f \mid \lim_{x \to \infty} f(x) = \infty\}$.

Rosenlicht computed in a series of papers the value groups and principal ranks of some Hardy fields, and found interesting applications of his computations to the theory of asymptotic functions. However, the Hardy fields he considered are of various, predominantly finite ranks. Our approach in this monograph is complementary to Rosenlicht's, since the exponential fields that we consider necessarily have infinite rank (cf. Corollary 1.23). In particular this is true for exponential Hardy fields, which we introduce now.

We call a Hardy field H an **exponential Hardy field** if it is real closed and satisfies the following conditions:

(EH1) $\mathbb{R}(x) \subset H$,
(EH2) if $f \in H$, then $\exp(f) \in H$,
(EH3) if $f \in H$, $f > 0$ then $\log(f) \in H$.

Here, $\exp(f)$ is just the germ of $(\exp \circ f)$, and similarly for $\log(f)$. The following easy observation subsumes exponential Hardy fields in the class of ordered exponential fields that we have been studying so far:

Lemma 6.15 *Let H be an exponential Hardy field. Then the map*
$$f \mapsto \exp(f)$$
is a (GA)-exponential on H.

Below, we denote by $\log_n$ (respectively, $\exp_n$) the n-th iterate of log (respectively, of exp). That is, for positive infinite elements $z \in H$, we set $\log_0(z) = z$ and $\log_{m+1}(z) = \log(\log_m(z))$ for all integers $m \geq 0$; note that every $\log_m(z)$ is again positive infinite. Similarly, we define $\exp_m(z)$ for every $z \in H$. (Note that in Chapter 3 we have denoted the n-th iterate of the logarithm and the exponential by ℓ^n and f^n, respectively. Here, we want to avoid the superscript n since, because of multiplication of germs in a Hardy field, there is danger of confusion with raising to the power n). We shall write sometimes for convenience $\log_{-m}$ for $\exp_m$ (with $m \geq 0$).

Corollary 1.23 implies that the principal rank of an exponential Hardy field is uncountable. Even without the use of Corollary 1.23, one immediately sees that the principal rank must be infinite. Indeed, because of (GA),
$$\ldots \log_n(x), \ldots, \log_2(x), \log(x), x, \exp(x), \exp_2(x), \ldots, \exp_n(x), \ldots$$
are all positive infinite and lie in distinct multiplicative classes.

Example 6.16 The field LE of **Logarithmico-Exponential functions** that was introduced by Hardy (cf. [HD2]) is an exponential Hardy field. It is the field of germs of compositions of semi-algebraic functions, exp and log.

This example is not isolated; the next theorem implies that *every* Hardy field can be enlarged to an exponential Hardy field (cf. [RO1]).

Theorem 6.17 *For any Hardy field k, there is a real closed Hardy field $H \supset k$ which contains the germ y of any differentiable real valued function (defined on a positive half line) that satisfies a differential equation*
$$y' = F(y)$$
for $F(Y) \in H(Y)$.

In particular, $\mathbb{R} \subset H$ and for each $a \in H$, H contains $\exp(a)$ and an antiderivative of a. Thus if $a > 0$, then $\log(a) \in H$.

Other interesting examples of Hardy fields arise from o-minimal expansions of the reals. Let us briefly introduce this notion here. Let $\mathcal{R} = (\mathbb{R}, +, -, \cdot, 0, 1, <, \mathcal{F})$ be an expansion of the ordered field of real numbers by a set of real valued functions

2. Hardy fields

$\mathcal{F}$ ($f \in \mathcal{F}$ is a function $f : \mathbb{R}^n \to \mathbb{R}$, for some $n \in \mathbb{N}$). A subset $S \subset \mathbb{R}^n$ is $\mathbb{R}$-**definable in** $\mathcal{R}$ if there exist an elementary formula $\varphi(x_1, \ldots, x_n, y_1, \ldots, y_m)$ in the language $L(\mathcal{R})$ and parameters $b_1, \ldots, b_m \in \mathbb{R}$ such that

$$S = \{(a_1, \ldots, a_n) \in \mathbb{R}^n \mid \varphi(a_1, \ldots, a_n, b_1, \ldots, b_m) \text{ holds in } \mathcal{R}\}.$$

Similarly, a function $g : \mathbb{R}^m \to \mathbb{R}^l$ is $\mathbb{R}$-definable if its graph is $\mathbb{R}$-definable. $\mathcal{R}$ is **o-minimal** if every $\mathbb{R}$-definable subset of $\mathbb{R}$ is a finite union of points and intervals.

In [D–M–M1], the following is shown: if $\mathcal{R}$ is an o-minimal expansion of $(\mathbb{R}, +, -, \cdot, 0, 1, <)$, then the set of germs of $\mathbb{R}$-definable functions (in one variable) forms a Hardy field. We shall denote this Hardy field by $H(\mathcal{R})$. In the next section we shall study Hardy fields arising from o-minimal expansions.

The following two notions are of crucial importance:
- An o-minimal expansion is said to be **polynomially bounded** if for every $\mathbb{R}$-definable function f (of one variable), there exists an $n \in \mathbb{N}$ such that ultimately $|f(x)| \leq x^n$.
- It is is said to be **exponentially bounded** if for every $\mathbb{R}$-definable function f (of one variable), there exists an $n \in \mathbb{N}$ such that ultimately $|f(x)| \leq \exp_n(x)$.

We immediately see that an o-minimal expansion $\mathcal{R}$ is polynomially bounded if and only if the multiplicative-class of x (cf. Chapter 3) is the largest amongst the multiplicative-classes of positive infinite elements in $H(\mathcal{R})$ (equivalently, the sequence $(x^n \mid n \in \mathbb{N})$ is cofinal in $H(\mathcal{R})$). This implies that the principal rank of $H(\mathcal{R})$ has a last element (cf. Theorem 3.9).

Similarly, an o-minimal expansion in which exp is definable is exponentially bounded if and only if the exponential-class of x (cf. Chapter 3) is the largest amongst the exponential-classes of positive infinite elements in $H(\mathcal{R})$ (equivalently, the sequence $(\exp_n(x) \mid n \in \mathbb{N})$ is cofinal in $H(\mathcal{R})$). This implies that the principal exponential rank of $H(\mathcal{R})$ has a last element (cf. Theorem 3.33).

Clearly, if $\mathcal{R}$ is a polynomially bounded o-minimal expansion of the reals, then exp is not $\mathbb{R}$-definable. Conversely if $\mathcal{R}$ is an o-minimal expansion of the reals in which exp is not $\mathbb{R}$-definable, then it is necessarily polynomially bounded (cf. [MI]).

Let f be a real-valued function defined on a positive half-line of the reals. We say that f is a **transexponential** function if it has transexponential growth, that is,

$$x \geq n^2 \Rightarrow f(x) > \exp_n(x) \quad (n \in \mathbb{N}).$$

Clearly, if $\mathcal{R}$ is an exponentially bounded o-minimal expansion of the reals, then no transexponential is $\mathbb{R}$-definable. In [BO], Hardy fields containing transexponential functions are constructed. However, it is unknown whether there exist o-minimal expansions of the reals which are *not* exponentially bounded.

We now make the following observation. We denote by g^{-1} the *compositional inverse* of a germ g.

Proposition 6.18 *Let H be a Hardy field containing $\mathbb{R}(x)$ and closed under compositions and compositional inverses for positive infinite germs. If the principal rank of H has a last element (in particular, if it is finite), then necessarily the rank of H is 1.*

Proof Let $f \in \mathbf{P}_H$ such that $[f]\dot{}$ is last. We first claim that there exists $n \in \mathbb{N}$ such that $f \leq x^n$. If not, then for all $n \in \mathbb{N}$ we have $x^n < f$. But this implies that for all $n \in \mathbb{N}$,
$$f^n < f \circ f .$$
Indeed, Let $n \in \mathbb{N}$. Since $x^n < f$, there exists $r \in \mathbb{R}$ (and we may assume $r > 1$) such that for $a \in \mathbb{R}$ with $a > r$ we have $a^n < f(a)$. In particular, for such a, $f(a) > r$. Thus $f(a)^n < f(f(a))$. Thus $f^n < f \circ f$. But this means that $f \ll f \circ f$, which contradicts the assumption on f, and the claim is established. Since on the other hand by assumption, there exists $n \in \mathbb{N}$ such that $x < f^n$, the claim shows that $x \sim f$.

Secondly, we claim that if $[x]\dot{}$ is last, then it is the only class. This will establish the theorem. Let g be positive infinite, that is $\lim_{x \to \infty} g(x) = \infty$. Assume for a contradiction that $g \ll x$. But this implies that for all $n \in \mathbb{N}$,
$$x^n < g^{-1} .$$
Indeed, Let $n \in \mathbb{N}$. Since $g^n < x$, there exists $r \in \mathbb{R}$ (and we may assume $r > 1$) such that for $a \in \mathbb{R}$ with $a > r$ we have $g(a)^n < a$. On the other hand, g is invertible, ultimately. So for b large enough, $g^{-1}(b) = a$ exists and $a > r$. Thus $g(g^{-1}(b))^n < g^{-1}(b)$. We have thus shown that $x \ll g^{-1}$, which contradicts the assumption on x and establishes our second claim. □

Corollary 6.19 *Let $\mathcal{R}$ be an o-minimal expansion of the reals. Then $\mathcal{R}$ is polynomially bounded if and only if the rank of $H(\mathcal{R})$ is equal to 1.*

Proof We only have to note that $H(\mathcal{R})$ satisfies the assumptions of the proposition. But this is clear since compositions and compositional inverses of $\mathbb{R}$-definable functions are again $\mathbb{R}$-definable. □

By similar arguments (replacing the function x by $\exp(x)$, and multiplicative equivalence by exponential equivalence), we have the exponential analogue of Proposition 6.18 and Corollary 6.19:

Proposition 6.20 *Let H be an exponential Hardy field (thus containing $\mathbb{R}(x)$) and closed under compositions and compositional inverses for positive infinite germs. If the principal exponential rank of H has a last element (in particular, if finite), then necessarily the exponential rank of H is 1.*

Corollary 6.21 *An o-minimal expansion of the reals $\mathcal{R}$ in which $\exp$ is $\mathbb{R}$-definable is exponentially bounded if and only if the exponential rank of $H(\mathcal{R})$ is equal to 1.*

We shall need the following notion. Let f and $g \neq 0$ be two real valued functions defined on positive half-lines of the reals. Recall that f is said to be **asymptotic** to g (and write $f \sim g$) if
$$\lim_{x \to \infty} \frac{f(x)}{g(x)} = 1 .$$
We note this useful lemma:

Lemma 6.22 *Let $f, g \neq 0$ be elements of the Hardy field H. Then f is asymptotic to g if and only if*
$$v(f - g) > v(g) .$$

Proof Clearly f is asymptotic to g if and only if $\lim_{x \to \infty} \frac{f(x)}{g(x)} - 1 = 0$. In other words $v(\frac{f}{g} - 1) > 0$, which is equivalent to $v(\frac{f-g}{g}) > 0$, and to $v(f - g) > v(g)$. □

Let H be an exponential Hardy field and $a \in \mathbf{P}_H$. Let $s \in \mathbb{Z}$. As in [RO3], we say that a has **level** s if there is $N \in \mathbb{N}$ such that
$$\log_{N+s}(a) \sim \log_N(x) \, .$$
H has levels or is **levelled** if every $a \in \mathbf{P}_H$ has a level.

3 Value groups

Let $\mathcal{P} = (\mathbb{R}, +, -, \cdot, 0, 1, <, \mathcal{F}_\pi)$ be a polynomially bounded o-minimal expansion of the reals by a collection of functions $\mathcal{F}_\pi$ consisting of $f : \mathbb{R}^m \to \mathbb{R}$ for various $m \in \mathbb{N}$. Examples of such expansions are: the expansion by restricted analytic functions $\mathbb{R}_{\mathrm{an}} = (\mathbb{R}, +, -, \cdot, 0, 1, <, \mathcal{F}_{\mathrm{an}})$ (cf. [D–M–M1] and Section 7 for more details), the expansion by convergent generalized power series (cf. [D–S1]), and the expansion by multisummable real power series (including the Gevrey functions, cf. [D–S2]). We denote by $T(\mathcal{P})$ the elementary theory of $(\mathbb{R}, +, -, \cdot, 0, 1, <, \mathcal{F}_\pi)$.

As in [D–S2], we assume for convenience that each 0-definable function (i.e., definable without real parameters) is the interpretation in $\mathcal{P}$ of a function symbol of the language $L(\mathcal{P})$.

Following [D–S2], we let the **field of exponents** of $\mathcal{P}$ be the set of exponents of the 0-definable power functions:
$$\kappa = \{r \in \mathbb{R} \mid \text{the function } x \mapsto x^r : (0, \infty) \to \mathbb{R} \text{ is 0-definable in } \mathcal{P}\} \, .$$
For example, if $\mathcal{F}_\pi = \mathcal{F}_{\mathrm{an}}$, then $\kappa = \mathbb{Q}$ (cf. [D–S2]). Note that the value group $v(M)$ of a model M of $T(\mathcal{P})$ has a natural κ-vector space structure:
$$rv(x) = v(x^r) \, .$$

We work in a model M of $T(\mathcal{P})$. Let $\mathcal{F} \subset \mathcal{F}_\pi$. Let F be an arbitrary subfield $F \subset M$. The real closure F^{r} of F can be taken to lie in M since M is real closed. We denote by F^{h} the henselization of (F, v). It can be taken to lie in M since by Lemma 6.10, (M, v) is henselian. We let $F^{\mathcal{F}}$ denote the smallest subfield of M which contains F and is closed under all functions of $\mathcal{F}$. Further, we let $F^{\mathrm{h}\mathcal{F}}$ denote the smallest subfield of M which is closed under all functions of $\mathcal{F}$, henselian for v and contains F. Similarly, we denote by $F^{\mathrm{r}\mathcal{F}}$ the smallest subfield of M which is real closed, closed under all functions of $\mathcal{F}$ and contains F. We will say that F is $\mathcal{F}$-**closed** if $F = F^{\mathcal{F}}$, and r$\mathcal{F}$-**closed** if $F = F^{\mathrm{r}\mathcal{F}}$. Note that $F^{\mathcal{F}} \subset F^{\mathrm{h}\mathcal{F}} \subset F^{\mathrm{r}\mathcal{F}}$.

We shall use the following result, which is a special case of the "valuation property" for polynomially bounded o-minimal expansions:

Lemma 6.23 *Let M and N be models of $T(\mathcal{P})$ with $M \subset N$. If $0 \neq x \in N$ such that $v(x) \notin v(M)$ then*
$$v(M(x)^{\mathrm{r}\mathcal{F}_\pi}) = v(M) + \kappa v(x) \, .$$

The valuation property was proved in full generality in [D–S2], Proposition 9.2. In the case of $\mathcal{F}_\pi = \mathcal{F}_{\mathrm{an}}$, it is the valuation property of restricted analytic functions (Corollary 3.7 of [D–M–M1]). Note that here we only use the case of the valuation property where the value group changes.

The following lemma is a consequence of the valuation property:

Lemma 6.24 *Assume that K is an $r\mathcal{F}_\pi$-closed subfield of M. Let I be some index set. Take $\{x_i \mid i \in I\}$ such that the values $\{v(x_i) \mid i \in I\}$ are κ-linearly independent over $v(K)$. Then*

$$v(K(x_i \mid i \in I)^{r\mathcal{F}_\pi}) = v(K) \oplus \bigoplus_{i \in I} \kappa v(x_i) \;. \tag{64}$$

In particular, if $\mathbb{R} \subset M$ and $x_i \in M$ are such that $v(x_i)$, $i \in I$, are κ-linearly independent, then

$$v(\mathbb{R}(x_i \mid i \in I)^{r\mathcal{F}_\pi}) = \bigoplus_{i \in I} \kappa v(x_i) \;.$$

Proof Clearly, it is sufficient to show the lemma for $I = \{1, \ldots, n\}$ finite.

Suppose we have already shown (64) for some $j < n$ in the place of n. Since the $v(x_i)$ are κ- independent over $v(K)$,

$$v(x_{j+1}) \notin v(K) \oplus \bigoplus_{i=1}^{j} \kappa v(x_i) = v(K(x_1, \ldots, x_j)^{r\mathcal{F}_\pi}) \;.$$

Lemma 6.23 shows that

$$\begin{aligned}
v(K(x_1, \ldots, x_{j+1})^{r\mathcal{F}_\pi}) &= v\left((K(x_1, \ldots, x_j)^{r\mathcal{F}_\pi})(x_{j+1})^{r\mathcal{F}_\pi}\right) \\
&= v(K(x_1, \ldots, x_j)^{r\mathcal{F}_\pi}) \oplus \kappa v(x_{j+1}) \\
&= v(K) \oplus \bigoplus_{i=1}^{j+1} \kappa v(x_i) \;.
\end{aligned}$$

By induction, we obtain the result for every n. $\square$

4 The Hardy field of a polynomially bounded + (exp) expansion

In this section, we consider $\mathcal{F}_\pi$ as in the previous section, but always assume in addition that the restricted functions $\{\exp |[0,1], \log |[1,e]\} \subset \mathcal{F}_\pi$ (we agree to set these functions to be identically 0 outside these closed intervals).

In [D–S2], it is shown that in this case, the expansion $(\mathcal{P}, \exp)$ obtained by expanding the structure $\mathcal{P}$ by the *unrestricted* exponential function, is again o-minimal, and has a model complete theory. Moreover, the Hardy field $H(\mathcal{P}, \exp)$ is a model of $T(\mathcal{P}, \exp)$, the elementary theory of $(\mathbb{R}, +, -, \cdot, 0, 1, <, \mathcal{F}_\pi, \exp)$.

We work in a model M of $T(\mathcal{P}, \exp)$. We assume that M properly contains $\mathbb{R}$, and take $x \in M$ to be a positive infinite element, that is, $x > \mathbb{R}$. For $\mathcal{F} \subset \mathcal{F}_\pi$ we are interested in the smallest subfield F of M containing $\mathbb{R}(x)$ which is $r\mathcal{F}$-closed and now in addition is

- exp-closed, i.e., $\exp(a) \in F$ for every $a \in F$,
- log-closed, i.e., $\log(a) \in F$ for every *positive* $a \in F$.

We will denote this closure of $\mathbb{R}(x)$ by $LE_\mathcal{F}(x)$. In the sequel, we shall give a precise description of this closure. We will apply our results to construct in particular the Hardy field $H(\mathcal{P}, \exp)$. Indeed, by [D–M–M1],

$$H(\mathcal{P}, \exp) = LE_{\mathcal{F}_\pi}(x) \tag{65}$$

(where x denotes the germ of the identity function).

Note that by the above condition on $\mathcal{F}_\pi$, if a subfield F of M is $\mathcal{F}_\pi$-closed, then $\exp \varepsilon \in F$ and $\log(1 + \varepsilon) \in F$ for every infinitesimal ε in F.

4. The Hardy field of a polynomially bounded + (exp) expansion

Lemma 6.25 (Closing under exp) *Let K be a log- and $r\mathcal{F}_\pi$-closed subfield of M, containing $\mathbb{R}$. If $a \in K$ such that $\exp(a) \notin K$, then $v(\exp(a)) \notin v(K)$.*

Proof Suppose for a contradiction that $v(\exp(a)) = v(z)$ for some $z \in K$, $z > 0$. Thus $\exp(a) = zu$ for some positive unit of M. Taking logs, we get $a - \log z = \log u$. Thus $\log u \in K$. Now u is a unit, thus $\log u$ is in the valuation ring, so $\log u = r + \varepsilon$ with $v(\varepsilon) > 0$, $r \in \mathbb{R}$. Thus $\varepsilon \in K$. We have:
$$u = \exp(\log u) = \exp(r)\exp(\varepsilon) .$$
Now $\exp(r) \in \mathbb{R}$, thus $\exp(r) \in K$. On the other hand, since K is $\mathcal{F}_\pi$-closed and $\{\exp|[0,1], \log|[1,e]\} \subset \mathcal{F}_\pi$, it follows that $\exp(\varepsilon) \in K$. Thus $u \in K$, so $\exp(a) = zu \in K$, a contradiction. □

Lemma 6.26 (Closing under log) *Assume that $K = \mathbb{R}(x_i \mid i \in I)^{r\mathcal{F}_\pi} \subset M$ such that*
1) the values $v(x_i)$, $i \in I$, are κ-independent,
2) for all $i \in I$, $x_i > 0$ and $\log(x_i) \in K$.
Then K is log-closed.

Proof Take $b \in K$, $b > 0$. By Lemma 6.24, there is a finite subset $I_0 \subset I$ and $r_i \in \kappa$ such that $v(b) = \sum_{i \in I_0} r_i v(x_i)$. So we can write $b = \prod_{i \in I_0} x_i^{r_i} \cdot r \cdot (1 + \varepsilon)$ with $r \in \mathbb{R}$ and $\varepsilon \in K$ such that $v(\varepsilon) > 0$. We have that $\log(1+\varepsilon) \in K$ since K is $\mathcal{F}_\pi$-closed. Moreover, $\log r \in \mathbb{R} \subset K$. Therefore,
$$\log(b) = \sum_{i \in I_0} r_i \log(x_i) + \log(r) + \log(1+\varepsilon) \in K .$$
□

The last two lemmas imply the following very useful result:

Corollary 6.27 (Closing under exp and log) *Assume that K is of the form*
$$\mathbb{R}(x_i \mid i \in I)^{r\mathcal{F}_\pi} \quad \text{log-closed, with } x_i > 0 \text{ and } v(x_i), \ i \in I, \ \kappa\text{-independent.} \tag{66}$$
Take any $a \in K$ such that $\exp(a) \notin K$. Then $v(\exp(a))$ is κ-independent over $v(K)$, and
$$v(K(\exp(a))^{r\mathcal{F}_\pi}) = v(K) \oplus \kappa\, v(\exp(a)) . \tag{67}$$
Moreover,
$$K(\exp(a))^{r\mathcal{F}_\pi} = \mathbb{R}(\exp(a), x_i \mid i \in I)^{r\mathcal{F}_\pi}$$
is again log-closed, and therefore of the form (66). It contains $\exp(b)$ whenever $b \in K(\exp(a))^{r\mathcal{F}_\pi}$ and $v(\exp(b))$ is κ-dependent over $v(K(\exp(a))^{r\mathcal{F}_\pi})$.

Proof Applying Lemma 6.25, we obtain that $v(\exp(a))$ is κ-independent over $v(K)$ and that $\exp(b) \in K(\exp(a))^{r\mathcal{F}_\pi}$ whenever $b \in K(\exp(a))^{r\mathcal{F}_\pi}$ and $v(\exp(b))$ is κ-dependent over $v(K(\exp(a))^{r\mathcal{F}_\pi})$. Equation (67) follows from Lemma 6.24. We infer from Lemma 6.26 that $K(\exp(a))^{r\mathcal{F}_\pi}$ is log-closed. □

Next, we show how to construct such log-closed fields K. **From now on, we always assume that $x \in M$ is a positive infinite element, i.e., $x > 0$ and $v(x) < 0$.**

Lemma 6.28 *The field*
$$\mathbb{R}(\log_m(x) \mid m \geq 0)^{r\mathcal{F}_\pi}$$
is log-closed. The convex hull of its value group in $v(M)$ is equal to the smallest convex subgroup containing $v(x)$.

Proof From Lemma 6.11 we know that for all $n \in \mathbb{N}$
$$v(x) < nv(\log(x)) < v(\log(x)) < \ldots < v(\log_{m-1}(x)) < nv(\log_m(x)) < \ldots < 0 \,. \tag{68}$$

Hence, the values $v(\log_m(x))$ are pairwise archimedean inequivalent. Since κ is archimedean, it follows that those values are κ-linearly independent. So it follows from Lemma 6.26 that $\mathbb{R}(\log_m(x) \mid m \geq 0)^{r\mathcal{F}_\pi}$ is log-closed.

From Lemma 6.24 we infer that
$$v(\mathbb{R}(\log_m(x) \mid m \geq 0)^{r\mathcal{F}_\pi}) = \bigoplus_{m \geq 0} \kappa\, v(\log_m(x)) \,.$$

Now (68) yields that this group is contained in the smallest convex subgroup of $v(M)$ which contains $v(x)$, in particular, its convex hull must be equal to the smallest convex subgroup containing $v(x)$. $\square$

The following corollary follows immediately from the lemma, together with the fact that the values $v(\log_m(x))$ are pairwise archimedean inequivalent, for distinct m.

Corollary 6.29 *Consider $K_0 = \mathbb{R}(\log_m(x) \mid m \geq 0)^{r\mathcal{F}_\pi}$. Then $v(K_0)$ is the lexicographic product*
$$v(\mathbb{R}(\log_m(x) \mid m \geq 0)^{r\mathcal{F}_\pi}) = \coprod_{m \geq 0} \kappa\, v(\log_m(x)) \,.$$
Thus the sequence $(v(\log_m(x)) \mid m \geq 0)$ is cofinal in $v(K_0)^{<0}$.

Next, we build up $LE_{\mathcal{F}_\pi}(x)$.

Theorem 6.30 *There exists $\{x_i \mid i \in I\} \subset LE_{\mathcal{F}_\pi}(x)$ such that*
1) *$\{v(x_i) \mid i \in I\}$ is a set of κ-linearly independent elements,*
2) *$LE_{\mathcal{F}_\pi}(x) = \mathbb{R}(x_i \mid i \in I)^{r\mathcal{F}_\pi}$.*

Furthermore, $\{x_i \mid i \in I\}$ can be chosen so that it contains $\{\log_m(x) \mid m \geq 0\}$.

Proof To get started, let K_0 be an *arbitrary* field of the form (66) such that $\mathbb{R}(x) \subset K_0 \subset LE_{\mathcal{F}_\pi}(x)$. For example, we can take $K_0 = \mathbb{R}(\log_m(x) \mid m \geq 0)^{r\mathcal{F}_\pi}$. By iterated application of Corollary 6.27, we construct a field $K_1 \subset LE_{\mathcal{F}_\pi}(x)$ of the form (66), such that $K_0 \subset K_1$ and K_1 contains $\exp(a)$ for every $a \in K_0$. Then we construct K_2 from K_1 in the same way as we constructed K_1 from K_0. We iterate to obtain fields $K_n \subset LE_{\mathcal{F}_\pi}(x)$, of the form (66). Their union
$$K_\infty := \bigcup_{n \in \mathbb{N}} K_n$$
contains $\mathbb{R}(x)$, is contained in $LE_{\mathcal{F}_\pi}(x)$ and is obviously $r\mathcal{F}_\pi$-closed, log-closed and of the form (66). It is also exp-closed. To see this, let $a \in K_\infty$. Then $a \in K_n$ for some n. Thus by construction, $\exp(a) \in K_{n+1} \subset K_\infty$. $\square$

This proves:

Corollary 6.31 *K_∞ is the uniquely determined smallest log-, exp- and $r\mathcal{F}_\pi$-closed subfield of M, and containing K_0. It is of the form (66).*

5. Exponential boundedness

We conclude:

Theorem 6.32 *The Hardy field $H(\mathcal{P}, \exp)$ is of the form (66). The elements x_i can be chosen so as to include the germ of the identity function x and $\log_m(x)$ for all $m \in \mathbb{N}$.*

We derive some crucial information about these closures $LE_{\mathcal{F}_\pi}(x)$ from our construction. We will use this in the next two sections to prove exponential-boundedness and the existence of levels in these Hardy fields.

Lemma 6.33 *Take $n \in \mathbb{N}$. If $a \in K_n$ with $v(a) < 0$ and $a > 0$, then $v(\log(a)) \in v(K_{n-1})$ and $v(\log_n(a)) \in v(K_0)$.*

Proof By the construction of K_n from K_{n-1}, there are elements $a_j \in K_{n-1}$, $j \in J$, such that $v(K_n) = v(K_{n-1}) \oplus \bigoplus_{j \in J} \kappa v(\exp(a_j))$. Hence, $a \in K_n$ can be written as

$$a = \prod_{j \in J_0} (\exp(a_j))^{q_j} \cdot c \cdot r \cdot (1 + \varepsilon)$$

with J_0 a finite subset of J, $q_j \in \kappa$, $c \in K_{n-1}$, $r \in \mathbb{R}$ and $\varepsilon \in K_n$ with $v(\varepsilon) > 0$. Then $\log(a) = \sum_{j \in J_0} q_j a_j + \log(c) + \log(r) + \log(1 + \varepsilon)$. Since $v(\log(a)) < 0$ by Lemma 6.11, but $v(\log(1 + \varepsilon)) > 0$, we find that

$$v(\log(a)) = v\left(\sum_{j \in J_0} q_j a_j + \log(c) + \log(r)\right) \in v(K_{n-1}) \, .$$

By induction it follows that $v(\log_n(a)) \in v(K_0)$. □

5 Exponential boundedness

We apply Lemma 6.33 to show that $LE_{\mathcal{F}_\pi}(x)$ has exponential rank 1. Note that exactly the same type of arguments were used in Chapter 5 to show that the exponential rank of the field $K_0 = \mathbb{R}((\mathbb{R}^{\Gamma_0}))$ with the prelogarithm $\log_\sigma$ determines the exponential rank of the exponential-logarithmic power series field $\mathbb{R}((\Gamma_0))^{EL}$.

Corollary 6.34 *The sequence*

$$(v(\log_m(x)) \mid m \geq 0)$$

is cofinal in $v(LE_{\mathcal{F}_\pi}(x))^{<0}$. Equivalently, the sequence eventually enters every non-zero convex subgroup H_w of $v(LE_{\mathcal{F}_\pi}(x))$.

Equivalently, the sequence $(\log_m(x) \mid m \geq 0)$ is coinitial among the positive infinite elements of $LE_{\mathcal{F}_\pi}(x)$. Equivalently, the sequence $(\log_m(x) \mid m \geq 0)$ eventually enters every convex valuation ring R_w of $LE_{\mathcal{F}_\pi}(x)$, with $w \neq v$.

Proof We choose to start our construction of $LE_{\mathcal{F}_\pi}(x)$ by setting

$$K_0 = \mathbb{R}(\log_m(x) \mid m \geq 0)^{\mathrm{r}\mathcal{F}_\pi}$$

in the proof of Theorem 6.30. Let H_w be a non-zero convex subgroup, and $\alpha < 0$, $\alpha \in H_w$. Let $a > 0$, $a \in K_\infty$ such that $v(a) = \alpha$. Thus for some n, $a \in K_n$ and by Lemma 6.33, $v(\log_n(a)) \in v(K_0)$. Now $\alpha = v(a) < v(\log_n(a)) < 0$. By Lemma 6.29 there is $m_0 \in \mathbb{N}$ such that $v(\log_n(a)) < v(\log_{m_0}(x)) < 0$. Thus $\alpha < v(\log_{m_0}(x)) < 0$, as required. □

From the above we immediately get:

Corollary 6.35 *The sequence $\exp_m x$, $m \geq 0$, is cofinal in $LE_{\mathcal{F}_\pi}(x)$.*

From the above, Proposition 6.20 and Corollary 6.21 we get:

Corollary 6.36 *The Hardy field $H(\mathcal{P}, \exp)$ has exponential rank 1. Thus the o-minimal expansion of the reals $(\mathcal{P}, \exp)$ is exponentially bounded.*

The exponential boundedness was shown for the case of $H(\mathbb{R}_{\mathrm{an}}, \exp) = LE_{\mathcal{F}_{\mathrm{an}}}(x)$ in [D–M–M2], and in general for $H(\mathcal{P}, \exp)$ in [D–S2], by other methods.

6 Levels

We shall establish the following:

Theorem 6.37 *The Hardy field $H(\mathcal{P}, \exp)$ has levels.*

Proof Let a be a positive infinite germ in $LE_{\mathcal{F}_\pi}(x)$. We shall compute explicitly the level of a. According to our construction, we write $LE_{\mathcal{F}_\pi}(x) = K_\infty$ with $K_0 = \mathbb{R}(\log_m(x) \mid m \geq 0)^{\mathrm{r}\mathcal{F}_\pi}$. By Lemma 6.33 there is some $n \in \mathbb{N}$ such that $v(\log_n(a)) \in v(K_0)$. Similarly as in the proof of Lemma 6.26, we write $\log_n(a) = \prod_{i \geq 0} (\log_i(x))^{q_i} \cdot r \cdot (1 + \varepsilon)$ with $q_i \in \kappa$, only finitely many of them non-zero, $r \in \mathbb{R}$ and $\varepsilon \in K$ such that $v(\varepsilon) > 0$. It follows that

$$\log_{n+1}(a) = \sum_{i \geq 0} q_i \log_{i+1}(x) + \log(r) + \log(1 + \varepsilon) .$$

Let i_0 be the smallest of all $i \geq 0$ for which $q_i \neq 0$. We have that $v(\log(r)) \geq 0$, $v(\log(1 + \varepsilon)) > 0$ and $v(\log_{i_0+1}(x)) < v(\log_{i+1}(x))$ for $i > i_0$. Thus, we can write

$$\log_{n+1}(a) = q_{i_0} \log_{i_0+1}(x) \cdot (1 + \varepsilon')$$

with $v(\varepsilon') > 0$. Then

$$\log_{n+2}(a) = \log(q_{i_0}) + \log_{i_0+2}(x) + \log(1 + \varepsilon') .$$

Again,

$$v(\log_{i_0+2}(x)) < 0 = v(\log(q_{i_0})) < v(\varepsilon') = v(\log(1 + \varepsilon')) .$$

Hence,

$$\begin{aligned} v(\log_{n+2}(a) - \log_{i_0+2}(x)) &= v(\log(q_{i_0}) + \log(1 + \varepsilon')) = v(\log(q_{i_0})) \\ &= 0 > v(\log_{i_0+2}(x)) . \end{aligned}$$

By Lemma 6.22, this shows that a has level $n - i_0$, and the theorem is proved. □

The existence of levels for the special case $H(\mathbb{R}_{\mathrm{an}}, \exp)$ was shown in [M–MI] by other methods.

To close this section, let us make the following important observations:

Remark 6.38 Let M be a model of $T(\mathcal{P}, \exp)$, and $\mathcal{F} \subset \mathcal{F}_\pi$. We wish to consider $\mathcal{F}$-closures of subfields of M in the place of $\mathcal{F}_\pi$-closures.

In fact, the results obtained above concerning $\mathcal{F}_\pi$-closures hold for $\mathcal{F}$-closures *provided* that $\{\exp|[0,1], \log|[1,e]\} \subset \mathcal{F}$ and $\mathcal{F}$ contains the 0-definable power functions $\{x^r \mid r \in \kappa\}$. Indeed, the key Lemma 6.24 holds already for any $\mathcal{F}$ which contains the 0-definable power functions. To see this, assume as in the lemma that K is an $r\mathcal{F}_\pi$-closed subfield of a model M of $T(\mathcal{P}, \exp)$ (for example, $K = \mathbb{R}$), and $x_i \in M$ with $v(x_i)$ $(i \in I)$ are κ-independent over $v(K)$.

Now observe that

$$v(K(x_i \mid i \in I)^{\mathrm{r}\mathcal{F}}) \subset v(K(x_i \mid i \in I)^{\mathrm{r}\mathcal{F}_\pi}) = v(K) \oplus \bigoplus_{i \in I} \kappa v(x_i) . \tag{69}$$

On the other hand, for every $r \in \kappa$ and every $i \in I$ we have $x_i^r \in K(x_i \mid i \in I)^{\mathcal{F}}$ by assumption on $\mathcal{F}$ (without loss of generality, we may assume that $x_i > 0$ for every $i \in I$). In particular, $rv(x_i) = v(x_i^r) \in v(K(x_i \mid i \in I)^{\mathcal{F}})$, which establishes the required result.

Similarly, all results of these sections hold for $\mathcal{F}$ which satisfy
$$\{\exp\!\mid\![0,1], \log\!\mid\![1,e]\} \subset \mathcal{F} \subset \mathcal{F}_{\mathrm{an}},$$
(see next section for the definition of $\mathcal{F}_{\mathrm{an}}$). Indeed, in that case the field of exponents is $\mathbb{Q}$. We compute as in the proof of Lemma 6.24, and verify that this lemma holds for any $\mathcal{F}$ already under the assumption that $\mathcal{F} \subset \mathcal{F}_{\mathrm{an}}$. Given an $r\mathcal{F}_{\mathrm{an}}$-closed subfield K of M (for example, $\mathbb{R}$), we have that
$$K(x_i \mid i \in I)^r \subset K(x_1, \ldots, x_n)^{r\mathcal{F}} \subset K(x_1, \ldots, x_n)^{r\mathcal{F}_{\mathrm{an}}}.$$
By Lemmas 6.3, 6.10 and 6.24 we get:
$$\begin{aligned} v(K(x_i \mid i \in I)^r) &= v(K) \oplus \bigoplus_{i \in I} \mathbb{Q}v(x_i) \subset v(K(x_i \mid i \in I)^{r\mathcal{F}}) \\ &\subset v(K(x_i \mid i \in I)^{r\mathcal{F}_{\mathrm{an}}}) = v(K) \oplus \bigoplus_{i \in I} \mathbb{Q}v(x_i) . \end{aligned}$$
So the required equality must hold.

Once Lemma 6.24 is established for $\mathcal{F}$-closures, Corollary 6.27, Lemma 6.28, Corollary 6.34 and Corollary 6.29 are proved exactly in the same way for $\mathcal{F}$-closures, provided that $\{\exp\!\mid\![0,1], \log\!\mid\![1,e]\} \subset \mathcal{F}$. We shall use these observations in the next sections.

7 The Crucial Lemma for models of T_{an}

In view of the facts stated in Remark 6.1, the following natural question arises: suppose we have additional structure on our ordered fields, to which extent can this structure be preserved under the embedding of the residue fields? We wish to answer this for the case where the additional structure is given by restricted analytic functions and exponentiation.

Let $\mathbb{R}_{\mathrm{an}}$ denote the expansion of the reals by all restricted analytic functions. This expansion is obtained as follows (cf. [D–M–M1] for details). For every $m \in \mathbb{N}$, we let $\mathbb{R}\{X_1, \ldots, X_m\}$ denote the ring of all real power series in $X_1, \ldots, X_m$ that converge in a neighborhood of the unit box $I^m = [-1, 1]^m$. For $f \in \mathbb{R}\{X_1, \ldots, X_m\}$ we define $\tilde{f} : \mathbb{R}^m \to \mathbb{R}$ by
$$\tilde{f}(x) = \begin{cases} 0 & \text{if } x \notin I^m \\ f(x) & \text{if } x \in I^m. \end{cases} \tag{70}$$
We call $\tilde{f}$ a **restricted analytic function**. Then $\mathbb{R}_{\mathrm{an}}$ is the structure
$$(\mathbb{R}, +, -, \cdot, 0, 1, <, \{\tilde{f} \mid f \in \mathbb{R}\{X_1, \ldots, X_m\}, \text{ for some } m \in \mathbb{N}\}) .$$
We shall denote it by $\mathbb{R}_{\mathrm{an}} = (\mathbb{R}, +, -, \cdot, 0, 1, <, \mathcal{F}_{\mathrm{an}})$. We let T_{an} denote the elementary theory of $\mathbb{R}_{\mathrm{an}}$. This theory is model complete, o-minimal and polynomially bounded (cf. [D]).

We summarize below a few facts about T_{an} that we will use. In [D–M–M1], the connection between the restricted analytic functions and convergent power series representing them is explained in detail ("convergent" means "convergent near 0").

Therefore, following [D–M–M2], we will identify (by abuse of notation) $\mathcal{F}_{\mathrm{an}}$ with the set of all convergent power series (in finitely many variables), thus representing all restricted analytic functions. We consider $\mathcal{F} \subset \mathcal{F}_{\mathrm{an}}$, a set of convergent power series. For instance, $\mathcal{F}_{LE}$ will denote the set consisting of the power series expansions of $\exp(x)$, $\log(1+x)$ and $\frac{1}{1+x}$. Let M be a model of T_{an}. From [D–M–M1] we know that a subfield $F \subset M$ containing $\mathbb{R}$ is closed under the restricted analytic functions (represented by $\mathcal{F}$) if and only if for every convergent power series $f(X_1, \ldots, X_n) \in \mathcal{F}$ and all *infinitesimals* $\varepsilon_1, \ldots, \varepsilon_n \in F$, the element $f(\varepsilon_1, \ldots, \varepsilon_n) \in M$ lies in F.

In [D–M–M1] it is also shown that every real closed subfield which is a substructure of a model of T_{an} is itself a model of T_{an}. Hence, if M is a model of T_{an} and F a subfield of M, then $F^{\mathrm{r}\mathcal{F}_{\mathrm{an}}}$ is a model of T_{an}. Moreover, for any model M of T_{an} we have that $\mathbb{R} \subset M$.

Throughout this section, let M be a model of T_{an}, and $\mathcal{F} \subset \mathcal{F}_{\mathrm{an}}$ an arbitrary set of convergent power series representing restricted analytic functions, closed under partial derivations, but not necessarily containing $\mathcal{F}_{LE}$. We will determine in this section the value groups and residue fields (with respect to an arbitrary convex valuation w) of the $\mathcal{F}$-closures of subfields of models of T_{an}.

Lemma 6.39 (The Crucial Lemma) *Let $\mathbb{R} \subset F \subset M$ and w a convex valuation on M. Assume that $Fw \subset F$ and that Fw is $\mathcal{F}$-closed. Assume further that the value group $w(F)$ is archimedean. Then either $F^{\mathcal{F}}$ is embeddable in the completion of (F, w) (and in particular, $w(F^{\mathcal{F}}) = w(F)$ and $F^{\mathcal{F}} w = Fw$), or there is some $y \in F^{\mathcal{F}}$, $y \neq 0$, such that $w(y) > w(F)$ (and in particular, $w(F^{\mathcal{F}})$ is not archimedean if $w(F) \neq 0$).*

Proof By Zorn's Lemma, we find a maximal subfield F_0 of $F^{\mathcal{F}}$ containing F and embeddable in the completion of (F, w). Suppose that $F^{\mathcal{F}}$ is not embeddable in the completion of (F, w). Then $F_0 \neq F^{\mathcal{F}}$, that is, F_0 is not $\mathcal{F}$-closed. So let $f(X_1, \ldots, X_k) \in \mathcal{F}$ and $a = (a_1, \ldots, a_k) \in F_0^k$ with $v(a_i) > 0$ such that $f(a) \in F^{\mathcal{F}} \setminus F_0$. We write $a_i = c_i + \varepsilon_i$ with $c_i \in F_0 w = Fw$ and $w(\varepsilon_i) > 0$; let $c = (c_1, \ldots, c_k)$. By the Taylor expansion, the following assertions hold (they are elementary sentences and thus hold in the T_{an}-model M): for all $m \in \mathbb{N}$,

$$\left| f(a_1, \ldots, a_k) - \sum_{\nu=(0,\ldots,0)}^{(m,\ldots,m)} \frac{\partial^\nu f}{\partial X^\nu}(c_1, \ldots, c_k) \frac{\varepsilon^\nu}{\nu!} \right| \leq |\varepsilon_1 \cdot \ldots \cdot \varepsilon_k|^m$$

(for $\nu = (\nu_1, \ldots, \nu_k) \in \mathbb{N}^k$, $\frac{\partial^\nu f}{\partial X^\nu}$ stands for $\frac{\partial^{\nu_1} \ldots \partial^{\nu_k} f}{\partial X_1^{\nu_1} \ldots \partial X_k^{\nu_k}}$, and $\nu!$ stands for $\nu_1! \cdot \ldots \cdot \nu_k!$). By convexity of w, it follows that for all $m \in \mathbb{N}$,

$$w\left(f(a_1, \ldots, a_k) - \sum_{\nu=(0,\ldots,0)}^{(m,\ldots,m)} \frac{\partial^\nu f}{\partial X^\nu}(c_1, \ldots, c_k) \frac{\varepsilon^\nu}{\nu!} \right) \geq m(w(\varepsilon_1) + \ldots + w(\varepsilon_k)).$$

Since $w(F_0)$ is archimedean and $w(\varepsilon_i) > 0$, the sequence $m(w(\varepsilon_1) + \ldots + w(\varepsilon_k))$, $m \in \mathbb{N}$, is cofinal in $w(F_0)$. This shows that the partial sums form a Cauchy sequence in (F_0, w), with limit $f(a)$. Note that since $\mathcal{F}$ is closed under partial derivatives and Fw is $\mathcal{F}$-closed, the coefficients $\frac{\partial^\nu f}{\partial X^\nu}(c_1, \ldots, c_k)$ lie in $Fw \subset F_0$. So the partial sums are indeed elements of F_0.

Suppose that the sequence has no limit in F_0. Then we can apply Lemma 6.8 to obtain that $F_0(f(a))$ is embeddable in the completion of (F_0, w) and hence also in the completion of (F, w). But this contradicts the maximality of F_0. Hence, there is some $b \in F_0$ which is also a limit of this sequence (observe that it is not necessarily a Cauchy sequence in (M, w)). Then by Lemma 6.5, $w(f(a) - b) > w(F_0)$. With $y := f(a) - b \ne 0$, we have found the desired element y which satisfies $w(y) > w(F)$. $\square$

Lemma 6.40 *Take $x_i \in M$ such that the values $v(x_i)$, $i \in I$ are rationally independent. Further, let w be any convex valuation. Assume that there is a subset $I_w \subset I$ such that $w(x_i) = 0$ for all $i \in I_w$ and that the values $w(x_i)$, $i \in I \setminus I_w$ are rationally independent. Then*

$$w(\mathbb{R}(x_i \mid i \in I)^{r\mathcal{F}}) = \bigoplus_{i \in I \setminus I_w} \mathbb{Q}w(x_i) \quad \text{and} \quad w(\mathbb{R}(x_i \mid i \in I)^{h\mathcal{F}}) = \bigoplus_{i \in I \setminus I_w} \mathbb{Z}w(x_i),$$

and

$$\mathbb{R}(x_i \mid i \in I)^{r\mathcal{F}}w = \mathbb{R}(x_i \mid i \in I_w)^{r\mathcal{F}} = (\mathbb{R}(x_i \mid i \in I)w)^{r\mathcal{F}}$$
$$\mathbb{R}(x_i \mid i \in I)^{h\mathcal{F}}w = \mathbb{R}(x_i \mid i \in I_w)^{h\mathcal{F}} = (\mathbb{R}(x_i \mid i \in I)w)^{h\mathcal{F}}.$$

In particular, since the values $v(x_i)$, $i \in I$, are rationally independent, we have

$$v(\mathbb{R}(x_i \mid i \in I)^{r\mathcal{F}}) = \bigoplus_{i \in I} \mathbb{Q}v(x_i) \quad \text{and} \quad v(\mathbb{R}(x_i \mid i \in I)^{h\mathcal{F}}) = \bigoplus_{i \in I} \mathbb{Z}v(x_i). \quad (71)$$

Proof We set $L := \mathbb{R}(x_i \mid i \in I)$ and $K := \mathbb{R}(x_i \mid i \in I_w)$. By Corollary 6.4, $v(L) = \bigoplus_{i \in I} \mathbb{Z}v(x_i)$, $v(K) = \bigoplus_{i \in I_w} \mathbb{Z}v(x_i)$, $w(L) = \bigoplus_{i \in I \setminus I_w} \mathbb{Z}w(x_i)$, $Lw = K$. From Lemma 6.24 and Remark 6.38 we infer that

$$v(L^{r\mathcal{F}}) = \bigoplus_{i \in I} \mathbb{Q}v(x_i) = \mathbb{Q} \otimes v(L)$$
$$v(K^{r\mathcal{F}}) = \bigoplus_{i \in I_w} \mathbb{Q}v(x_i) = \mathbb{Q} \otimes v(K).$$

The former implies that $w(L^{r\mathcal{F}}) = \mathbb{Q} \otimes w(L)$, which is our assertion on the value groups for the r$\mathcal{F}$-closure.

We prove the assertions of our lemma for the h$\mathcal{F}$-closure. The proof for the residue field of the r$\mathcal{F}$-closure is analogous. If our assertions are not true, then there is some $b \in L^{h\mathcal{F}}$ such that $w(b) \notin \bigoplus_{i \in I \setminus I_w} \mathbb{Z}w(x_i)$ or $bw \notin K^{h\mathcal{F}}$. But b is already contained in some finitely generated subextension of $L^{h\mathcal{F}} | \mathbb{R}$. This in turn is contained in some subfield $\mathbb{R}(x_1, \ldots, x_n)^{h\mathcal{F}} \subset L^{h\mathcal{F}}$, where $x_1, \ldots, x_n$ are suitably chosen from the x_i's. So we see that it suffices to prove our lemma in the case of a finite set $I = \{1, \ldots, n\}$.

Since $v(K)$ is contained in the convex subgroup H_w associated with w, we find that also $v(K^{r\mathcal{F}}) = \mathbb{Q} \otimes v(K) \subset H_w$. That is, w is trivial on $K^{r\mathcal{F}}$ and thus also on $K^{h\mathcal{F}}$. Therefore, $K^{h\mathcal{F}} \subset L^{h\mathcal{F}}w$. We will show that equality holds.

First assume that $w(L)$ is archimedean. Then also $w(L^{r\mathcal{F}}) = \mathbb{Q} \otimes w(L)$ is archimedean, and so is $w(L^{h\mathcal{F}}) \subset w(L^{r\mathcal{F}})$. Set $F := K^{h\mathcal{F}}(x_i \mid i \in I \setminus I_w)$. Then $L^{h\mathcal{F}} = K^{h\mathcal{F}}(x_i \mid i \in I \setminus I_w)^{h\mathcal{F}} = F^{h\mathcal{F}}$, and by Lemma 6.3, $Fw = K^{h\mathcal{F}}$ and $w(F) = w(L)$. By Zorn's Lemma, we find a maximal subfield F_0 of $F^{h\mathcal{F}}$ containing F and embeddable in the completion of (F, w). Since $w(F_0) = w(F)$ is archimedean and $F_0 w = Fw$ is $\mathcal{F}$-closed, we can apply Lemma 6.39 to see that F_0 is $\mathcal{F}$-closed. From Lemma 6.9 we infer that F_0 must be equal to its henselization, i.e., it is henselian. Therefore, $F_0 = F^{h\mathcal{F}} = L^{h\mathcal{F}}$, showing that

$$w(\mathbb{R}(x_i \mid i \in I)^{h\mathcal{F}}) = w(L^{h\mathcal{F}}) = w(F) = \bigoplus_{i \in I \setminus I_w} \mathbb{Z}w(x_i)$$

and
$$\mathbb{R}(x_i \mid i \in I)^{\mathrm{h}\mathcal{F}} w = L^{\mathrm{h}\mathcal{F}} w = Fw = K^{\mathrm{h}\mathcal{F}}.$$

(For the r$\mathcal{F}$-closure, one takes F_0 to be a maximal subfield of $F^{\mathrm{r}\mathcal{F}}$ containing F^{r} and embeddable in the completion of (F^{r}, w), and uses Lemma 6.10 in the place of Lemma 6.9.)

Now let $w(L)$ be non-archimedean. Since it is finitely generated, it has finite rank. So we can proceed by induction on the rank. Let C be the largest proper convex subgroup of $w(L)$. Since C is finitely generated, we can choose $y_1, \ldots, y_\ell \in L$ such that the values $w(y_1), \ldots, w(y_\ell)$ form a set of rationally independent generators of C. We take w' to be a convex valuation on M whose restriction to L is the valuation associated with C. Since $w(L)$ is finitely generated, we can choose $y_{\ell+1}, \ldots, y_m \in L$ such that the values $w'(y_i) = w(y_i) + C$, $\ell < i \leq m$, form a set of rationally independent generators of $w(L)/C$. Then $w(L) = \bigoplus_{1 \leq i \leq m} \mathbb{Z}w(y_i)$, and

$$\begin{aligned} m &= \ell + (m - \ell) = \dim_{\mathbb{Q}} \mathbb{Q} \otimes (w(L)/C) + \dim_{\mathbb{Q}} \mathbb{Q} \otimes C \\ &= \dim_{\mathbb{Q}} \mathbb{Q} \otimes w(L) = |I \setminus I_w| = \operatorname{trdeg} L|K. \end{aligned}$$

Since the values $w(y_1), \ldots, w(y_m)$ are rationally independent over $w(K) = \{0\}$, the elements $y_1, \ldots, y_m$ are algebraically independent over K. Consequently, L is algebraic over $K(y_1, \ldots, y_m)$. By our choice of the y_i, $w(L) = w(K(y_1, \ldots, y_m))$. By Lemma 6.3, $Lw = K = K(y_1, \ldots, y_m)w$. Thus, the extension is immediate. Now Lemma 6.7 shows that $K(y_1, \ldots, y_m)^{\mathrm{h}} = L^{\mathrm{h}}$. This implies that $L^{\mathrm{h}\mathcal{F}} = K(y_1, \ldots, y_m)^{\mathrm{h}\mathcal{F}}$, $K(y_1, \ldots, y_m)^{\mathrm{r}} = L^{\mathrm{r}}$, and $K(y_1, \ldots, y_m)^{\mathrm{r}\mathcal{F}} = L^{\mathrm{r}\mathcal{F}}$.

The rank of $w(K(y_1, \ldots, y_\ell)) = C$ is smaller than that of $w(L)$. Hence by induction hypothesis,

$$w(K(y_1, \ldots, y_\ell)^{\mathrm{h}\mathcal{F}}) = \bigoplus_{1 \leq i \leq \ell} \mathbb{Z}w(y_i) \quad \text{and} \quad K(y_1, \ldots, y_\ell)^{\mathrm{h}\mathcal{F}} w = K^{\mathrm{h}\mathcal{F}}.$$

On the other hand, the value group $w'(K(y_1, \ldots, y_m)) = w(L)/C$ is archimedean since C was chosen to be the largest proper convex subgroup of $w(L)$. By our choice of the elements y_i, $w'(y_i) = 0$ for $1 \leq i \leq \ell$, and the values $w'(y_{\ell+1}), \ldots, w'(y_m)$ are rationally independent. Thus, we can replace w by w' and apply the assertion of our lemma, which is already proved in the archimedean case, to deduce that

$$\begin{aligned} w'(K(y_1, \ldots, y_m)^{\mathrm{h}\mathcal{F}}) &= \bigoplus_{\ell < i \leq m} \mathbb{Z}w'(y_i) \\ K(y_1, \ldots, y_m)^{\mathrm{h}\mathcal{F}} w' &= K(y_1, \ldots, y_\ell)^{\mathrm{h}\mathcal{F}}. \end{aligned}$$

Since the values $w'(y_i) = w(y_i) + C$, $\ell < i \leq m$, are rationally independent, the values $w(y_i)$, $\ell < i \leq m$, are rationally independent over $C = \bigoplus_{1 \leq i \leq \ell} \mathbb{Z}w(y_i) = w(K(y_1, \ldots, y_\ell)^{\mathrm{h}\mathcal{F}})$. It follows that

$$\begin{aligned} w(\mathbb{R}(x_1, \ldots, x_n)^{\mathrm{h}\mathcal{F}}) &= w(K(y_1, \ldots, y_m)^{\mathrm{h}\mathcal{F}}) \\ &= w(K(y_1, \ldots, y_\ell)^{\mathrm{h}\mathcal{F}}) \oplus \bigoplus_{\ell < i \leq m} \mathbb{Z}w(y_i) \\ &= \bigoplus_{1 \leq i \leq \ell} \mathbb{Z}w(y_i) \oplus \bigoplus_{\ell < i \leq m} \mathbb{Z}w(y_i) \\ &= \bigoplus_{1 \leq i \leq m} \mathbb{Z}w(y_i) = w(L) = \bigoplus_{i \in I \setminus I_w} \mathbb{Z}w(x_i) \end{aligned}$$

and that
$$\mathbb{R}(x_1,\ldots,x_n)^{\mathrm{h}\mathcal{F}}w = K(y_1,\ldots,y_m)^{\mathrm{h}\mathcal{F}}w = (K(y_1,\ldots,y_m)^{\mathrm{h}\mathcal{F}}w')w$$
$$= K(y_1,\ldots,y_\ell)^{\mathrm{h}\mathcal{F}}w = K^{\mathrm{h}\mathcal{F}} = \mathbb{R}(x_i \mid i \in I_w)^{\mathrm{h}\mathcal{F}}.$$
□

Let us note that the result of this lemma remains true if the henselization with respect to v is replaced by the henselization with respect to any convex valuation.

For use in Section 10, we add the following lemma:

Lemma 6.41 *Let $x_i \in M$ such that $x_i > 0$ and the values $v(x_i)$, $i \in I$ are rationally independent. Further, let $x_i^{1/k}$ denote the unique positive real k-th root of x_i. Then*
$$\mathbb{R}(x_i \mid i \in I)^{\mathrm{r}\mathcal{F}} = \bigcup_{I_0 \subset I \text{ finite}} \bigcup_{k \in \mathbb{N}} \mathbb{R}(x_i^{1/k} \mid i \in I_0)^{\mathrm{h}\mathcal{F}} \tag{72}$$

with
$$v(\mathbb{R}(x_i^{1/k} \mid i \in I_0)^{\mathrm{h}\mathcal{F}}) = \bigoplus_{i \in I_0} \mathbb{Z} \frac{v(x_i)}{k},$$

a finitely generated group.

Proof The assertion for the value group follows from (71). Let U denote the union on the right-hand side of (72). Every field in the union is henselian, so U is henselian. The value group $v(U)$ is divisible and the residue field $Uv = \mathbb{R}$ is real closed. Hence by Lemma 6.10, U is real closed. By construction, U is also $\mathcal{F}$-closed. Since all $x_i^{1/k}$ are in the real closed field $\mathbb{R}(x_i \mid i \in I)^{\mathrm{r}\mathcal{F}}$, we find that U is contained in $\mathbb{R}(x_i \mid i \in I)^{\mathrm{r}\mathcal{F}}$. Since this field is the smallest real closed and $\mathcal{F}$-closed field containing $\mathbb{R}(x_i \mid i \in I)$, it follows that $U = \mathbb{R}(x_i \mid i \in I)^{\mathrm{r}\mathcal{F}}$. □

8 Residue fields of $\mathcal{F}$-exp-log-closures

In [D–M–M1] it is shown that the expansion $(\mathbb{R}_{\mathrm{an}},\exp)$ obtained by adding the *unrestricted* exponential function to the polynomially bounded expansion $\mathbb{R}_{\mathrm{an}}$ is again o-minimal. Moreover, the Hardy field $H(\mathbb{R}_{\mathrm{an}},\exp)$ is a model of the elementary theory of $(\mathbb{R}_{\mathrm{an}},\exp)$.

Throughout the next sections, we work in a model M of the elementary theory $T_{\mathrm{an}}(\exp)$ of $(\mathbb{R}_{\mathrm{an}},\exp)$. We assume that M properly contains $\mathbb{R}$, and take $x \in M$ to be a positive infinite element, that is, $x > \mathbb{R}$. We let $\mathcal{F} \subset \mathcal{F}_{\mathrm{an}}$ be a set of convergent power series representing restricted analytic functions, closed under partial derivations, and containing $\mathcal{F}_{LE}$. Hence, if $F \subset M$ is $\mathcal{F}$-closed, then $\exp \varepsilon \in F$ and $\log(1+\varepsilon) \in F$ for every infinitesimal ε in F.

Let $\mathcal{F}$ be given and $x \in M$ be positive infinite. In this section, we build up $LE_\mathcal{F}(x)$ and its residue fields from $\mathbb{R}(x)$. From this construction and the results of Section 7, we derive Theorem 6.44 (embedding of the residue fields with structure). We will build up the field $LE_\mathcal{F}(x)$ and its residue fields for an *arbitrary convex valuation w* on M. We allow w to be the trivial valuation, denoted by u.

In particular, since the Hardy field $H(\mathbb{R}_{\mathrm{an}},\exp)$ is a model of $T_{\mathrm{an}}(\exp)$, and since by [D–M–M1] we know that,
$$H(\mathbb{R}_{\mathrm{an}},\exp) = LE_{\mathcal{F}_{\mathrm{an}}}(x)$$

(where x denotes the germ of the identity function), our results will apply to $M = H(\mathbb{R}_{an}, \exp)$. We will use this description of the residue fields of $H(\mathbb{R}_{an}, \exp)$ to solve the Hardy problem (Section 9), and prove undefinability results (Section 10) *without* using the arguments about truncation closed embeddings which were used in [D–M–M2].

Let w be a convex valuation on M and H_w its associated convex subgroup of $v(M)$, R_w its valuation ring, $\mathcal{U}_w$ its group of units. For every subfield K of R_w, its multiplicative group $K^\times$ is contained in the multiplicative group $\mathcal{U}_w$ of all units of R_w. We will say that K is **relatively exp-closed in** $\mathcal{U}_w$ if $a \in K$ and $\exp(a) \in \mathcal{U}_w$ implies that $\exp(a) \in K$. For example, $\mathbb{R}$ is relatively exp-closed in $\mathcal{U}_w$ for every convex valuation w of M.

We prove the analogue of Lemma 6.25, now for arbitrary convex valuations w.

Lemma 6.42 *Let K be a log- and $r\mathcal{F}$-closed subfield of M, containing $\mathbb{R}$. Let w be a convex valuation of M. Assume that the residue field Kw is a subfield of $R_w \cap K$, relatively exp-closed in $\mathcal{U}_w$. Take any $a \in K$ such that $\exp(a) \notin K$. Then $w(\exp(a))$ is rationally independent over $w(K)$.*

Proof Suppose that $w(\exp(a))$ is not rationally independent over $w(K)$. Since $w(K)$ is divisible by Lemma 6.10, it follows that $w(\exp(a)) = w(b) \in w(K)$ for some positive $b \in K$. Then $w(\frac{\exp(a)}{b}) = 0$ and by Lemma 6.11,

$$w(a - \log(b)) = w\left(\log\left(\frac{\exp(a)}{b}\right)\right) \geq 0 \,.$$

Since K is log-closed, $\log(b) \in K$. Hence, there is $c \in Kw$ such that

$$w(a - \log(b) - c) > 0 \,.$$

By Lemma 6.11, this shows that

$$w\left(\frac{\exp(a)}{b \exp(c)}\right) = w(\exp(a - \log(b) - c)) = 0 \,.$$

In particular, we find that $w(\exp(c)) = w(\frac{\exp(a)}{b}) = 0$, that is, $\exp(c) \in \mathcal{U}_w$. By assumption on Kw, $\exp(c) \in Kw \subset K$.

By convexity of w, $w(a - \log(b) - c) > 0$ yields that $v(a - \log(b) - c) > 0$. Therefore, $\exp(a - \log(b) - c) \in K^{r\mathcal{F}} = K$, showing that

$$\exp(a) = \exp(a - \log(b) - c) \cdot b \cdot \exp(c) \in K \,.$$

We conclude: if $\exp(a) \notin K$, then $w(\exp(a))$ is rationally independent over $w(K)$.
□

Next, we show how to construct such log-closed fields K. **From now on, we always assume that $x \in M$ is a positive infinite element, i.e., $x > 0$ and $v(x) < 0$.** Recall that by Lemma 6.28 and Remark 6.38, the field

$$\mathbb{R}(\log_m(x) \mid m \geq 0)^{r\mathcal{F}}$$

is log-closed, for *any* positive infinite x. Now we prove:

Lemma 6.43 *Let $x \in M$ be positive infinite, and $w \neq v$ be a convex valuation on M. Then there exists an $m_0 \in \mathbb{N}$ such that $w(\log_{m_0}(x)) = 0$. Thus the log-closed field*

$$\mathbb{R}(\log_m(x) \mid m \geq m_0)^{r\mathcal{F}}$$

8. Residue fields of $\mathcal{F}$-exp-log-closures

lies in $R_w \cap LE_{\mathcal{F}}(x)$. It is of the form (66).

Proof First note that by Lemma 6.28 and Remark 6.38 we get that the field $K_0 := \mathbb{R}(\log_m(x) \mid m \geq m_0)^{\mathrm{r}\mathcal{F}}$ is log-closed, for any m_0 (since $\log_{m_0}(x)$ is again positive infinite).

Now by Corollary 6.34 and Remark 6.38, there exists an $m_0 \in \mathbb{N}$ such that $w(\log_{m_0}(x)) \geq 0$. On the other hand, $w(\log_{m_0}(x)) \leq 0$ since $\log_{m_0}(x)$ is positive infinite. So $w(\log_{m_0}(x)) = 0$. Thus $w(\log_m(x)) = 0$ for all $m \geq m_0$. Now by Corollary 6.29 and Remark 6.38, $v(K_0)$ is the lexicographic product

$$v(\mathbb{R}(\log_m(x) \mid m \geq m_0)^{\mathrm{r}\mathcal{F}}) = \coprod_{m \geq m_0} \mathbb{Q}\, v(\log_m(x)).$$

Thus $w(z) = 0$ for every $z \in K_0$, which establishes the required result. $\square$

We shall now prove the main theorem of this section:

Theorem 6.44 (Embedding of the residue field with structure) *Let w be an arbitrary convex valuation on M and $x \in M$ be positive infinite. Denote the valuation ring by R_w. Then there exists a real closed subfield*

$$K \subset R_w \cap LE_{\mathcal{F}}(x)$$

which is log-closed and $\mathcal{F}$-closed, relatively exp-closed in $\mathcal{U}_w$ and satisfies

$$K = LE_{\mathcal{F}}(x)w.$$

If $w = v$, then $K = \mathbb{R}$. Otherwise, if w is not the natural valuation, then there is some integer $m_0 \geq 0$ such that K can be chosen to be the uniquely determined smallest subfield of $R_w \cap LE_{\mathcal{F}}(x)$ which is real closed, log- and $\mathcal{F}$-closed, relatively exp-closed in $\mathcal{U}_w$ and contains $\mathbb{R}(\log_{m_0}(x))$. If $w(x) = 0$, then we can choose $m_0 = 0$, so that K contains x.

The assertion of the theorem for $w = v$ is clear. So assume that $w \neq v$. Using Lemma 6.43 we shall now construct a special subfield K_∞^w of R_w (which will turn out to be the residue field). The idea is similar to the one used in the proof of Theorem 6.30, except that now, we want to construct a field *within* a given valuation ring, and we want to *relatively* close under exp. Below, denote by H_w the convex subgroup associated to w.

To get started, let $K_0^w \subset R_w \cap LE_{\mathcal{F}}(x)$ be *any* field of the form (66). For example, if $w(\log_{m_0}(x)) = 0$, then we can take $K_0^w = \mathbb{R}(\log_m(x) \mid m \geq m_0)^{\mathrm{r}\mathcal{F}}$. We have that $v(K_0^w) \subset H_w$.

Now we construct K_1^w as follows. Assume that $a \in K_0^w$ such that $\exp(a) \notin K_0^w$, but $v(\exp(a)) \in H_w$. Then by Corollary 6.27 and Remark 6.38, $K_0^w(\exp(a))^{\mathrm{r}\mathcal{F}}$ is again of the form (66), with $v(K_0^w(\exp(a))^{\mathrm{r}\mathcal{F}}) = v(K_0^w) \oplus \mathbb{Q}\,v(\exp(a)) \subset H_w$. The latter shows that it is again a subfield of $R_w \cap LE_{\mathcal{F}}(x)$. We repeat this procedure until we arrive at a field $K_1^w \subset R_w \cap LE_{\mathcal{F}}(x)$ of the form (66), which contains $\exp(a)$ for every $a \in K_0^w$ such that $\exp(a) \in \mathcal{U}_w$. Then we construct K_2^w from K_1^w in the same way as we constructed K_1^w from K_0^w. We iterate to obtain fields $K_n^w \subset R_w \cap LE_{\mathcal{F}}(x)$, of the form (66). Their union

$$K_\infty^w := \bigcup_{n \in \mathbb{N}} K_n^w \subset R_w \cap LE_{\mathcal{F}}(x)$$

is log-closed and $\mathrm{r}\mathcal{F}$-closed and of the form (66). It is also relatively exp-closed in $\mathcal{U}_w$. To see this, let $a \in K_\infty^w$. Then $a \in K_n^w$ for some n. If $\exp(a) \in \mathcal{U}_w$, then by construction, $\exp(a) \in K_{n+1}^w$. On the other hand, every other log- and r$\mathcal{F}$-closed subfield of $R_w \cap LE_{\mathcal{F}}(x)$, relatively exp-closed in $\mathcal{U}_w$ and containing K_0^w, must also contain K_∞^w. This proves:

Lemma 6.45 K_∞^w is the uniquely determined smallest log- and r$\mathcal{F}$-closed subfield of $R_w \cap LE_{\mathcal{F}}(x)$, relatively exp-closed in $\mathcal{U}_w$ and containing K_0^w. It is of the form (66).

If $w(\log_{m_0}(x)) = 0$ and we start our construction from
$$K_0^w = \mathbb{R}(\log_m(x) \mid m \geq m_0)^{\mathrm{r}\mathcal{F}},$$
then K_∞^w will be the uniquely determined smallest log- and r$\mathcal{F}$-closed subfield of R_w, relatively exp-closed in $\mathcal{U}_w$ and containing $\mathbb{R}(\log_{m_0}(x))$. We denote it by
$$LE_{\mathcal{F}}^w(\log_{m_0}(x)).$$

Take $w = u$ to be the trivial valuation on M. Then $R_u = M$ and $H_u = v(M)$. In this case, $u(\log_0(x)) = u(x) = 0$ and $LE_{\mathcal{F}}^u(x)$ is exp-closed and contains x. Therefore,
$$LE_{\mathcal{F}}^u(x) = LE_{\mathcal{F}}(x).$$

This proves:

Theorem 6.46 $LE_{\mathcal{F}}(x)$ is of the form (66). The elements x_i can be chosen so as to include x and $\log_m(x)$ for all $m \in \mathbb{N}$.

Let m_0 be such that $w(\log_{m_0}(x)) = 0$. We want to show that
$$LE_{\mathcal{F}}(x)w = LE_{\mathcal{F}}^w(\log_{m_0}(x)). \tag{73}$$

By our construction, $LE_{\mathcal{F}}^w(\log_{m_0}(x)) \subset LE_{\mathcal{F}}(x) \cap R_w$ and is of the form (66). So we may rerun our construction of $LE_{\mathcal{F}}(x)$ starting with
$$K_0^u = LE_{\mathcal{F}}^w(\log_{m_0}(x)).$$

(Indeed, we will have at the end of the construction that K_∞^u is relatively exp-closed in $R_u = M$, that is, is exp-closed. Thus $x \in K_\infty^u$ and $K_\infty^u = LE_{\mathcal{F}}(x)$).

We note that K_0^u is of the form $\mathbb{R}(x_i \mid i \in I_w)^{\mathrm{r}\mathcal{F}}$, where the x_i, $i \in I_w$, are obtained from the above construction (and thus, their values $v(x_i) \in H_w$ are rationally independent). Since $K_0^u \subset R_w$, we have that $K_0^u w = K_0^u$ and that $w(x_i) = 0$ for $i \in I_w$. Suppose that while building up $LE_{\mathcal{F}}(x)$ from this field by the above construction, we have reached a field K of the form $\mathbb{R}(x_i \mid i \in I)^{\mathrm{r}\mathcal{F}}$ with $Kw = K_0^u$, $I_w \subset I$ and such that the values $w(x_i)$, $i \in I \setminus I_w$, are rationally independent. If $a \in K$, but $\exp a \notin K$, then Lemma 6.42 shows that $w(\exp(a))$ is rationally independent over $w(K)$. So the values $w(\exp(a)), w(x_i)$, $i \in I \setminus I_w$, are rationally independent, and Lemma 6.40 shows that
$$K(\exp(a))^{\mathrm{r}\mathcal{F}} w = \mathbb{R}(\exp(a), x_i \mid i \in I)^{\mathrm{r}\mathcal{F}} w = \mathbb{R}(x_i \mid i \in I_w)^{\mathrm{r}\mathcal{F}} = K_0^u.$$
Hence, $K(\exp(a))^{\mathrm{r}\mathcal{F}}$ is again of the same form as K. By induction, it follows that
$$LE_{\mathcal{F}}(x)w = LE_{\mathcal{F}}^w(\log_{m_0}(x)),$$
as required. Note that equation (73) holds with *every* natural number m_0 for which $w(\log_{m_0}(x)) = 0$. Thus we have established Theorem 6.44.

We derive the following important fact (which will play a crucial role in the solution of Hardy's problem).

Corollary 6.47 *Suppose that $\mathcal{F}_1 \subset \mathcal{F}_2$ are sets of convergent power series, closed under partial derivations and containing $\mathcal{F}_{LE}$. Let $w \neq v$ be a convex valuation on M. Let m_0 be such that $w(\log_{m_0}(x)) = 0$. Then*

$$LE_{\mathcal{F}_1}^w(\log_{m_0}(x)) \subset LE_{\mathcal{F}_2}^w(\log_{m_0}(x)).$$

Thus for every convex valuation w we have:

$$LE_{\mathcal{F}_1}(x)w \subset LE_{\mathcal{F}_2}(x)w.$$

Proof If $\mathcal{F}_1 \subset \mathcal{F}_2$, then the smallest log- and r$\mathcal{F}_1$-closed subfield of R_w, relatively exp-closed in $\mathcal{U}_w$ and containing $\mathbb{R}(\log_{m_0}(x))$, is contained in the smallest log- and r$\mathcal{F}_2$-closed subfield of R_w, relatively exp-closed in $\mathcal{U}_w$ and containing $\mathbb{R}(\log_{m_0}(x))$. □

Remark 6.48 Although we are explicitly treating only the case of $\mathcal{F}$ a set of restricted analytic functions, Theorem 6.44 and some other results on the valuation theoretical properties of $LE_{\mathcal{F}}(x)$ hold more generally. We may replace T_{an} by the theory of an arbitrary polynomially bounded o-minimal expansion of the reals, and take $\mathcal{F}$ to be any set of definable functions (such that $\mathcal{F}_{LE} \subset \mathcal{F}$) for which an analogue of our Crucial Lemma 6.39 can be proved. For example, Lemma 6.39 and its proof also hold if $\mathcal{F}$ is a set of Gevrey functions, closed under partial derivatives. Another interesting example is obtained if one replaces T_{an} by the theory of the reals with convergent generalized power series. In [D–S1], this theory is shown to be model complete and o-minimal. Take $\mathcal{F}$ to be a set of generalized power series for which the exponents of each variable form a sequence cofinal in $\mathbb{R}$ (indexed by the natural numbers). Then an analogue of Lemma 6.39 can be proved if $\mathcal{F}$ is closed under formal derivatives in the sense of [D–S1]. Although the condition on the exponents is quite restrictive, it holds for the presently known applications of interest. In particular, the function $\zeta(-\log(x)) = \sum_{n=1}^{\infty} x^{\log n}$ on $[0, e^{-2}]$ (with ζ the Riemann ζ-function) satisfies the condition.

In the next sections, we show that "truncation closed embeddings" of Hardy fields in logarithmic-exponential power series fields are not needed to derive the main results of [D–M–M2].

9 A truncation free solution to the Hardy problem

Recall that $H(\mathbb{R}_{\mathrm{an}}, \exp)$ denotes the field of the germs at $+\infty$ of all functions on $\mathbb{R}$ which are definable in $(\mathbb{R}_{\mathrm{an}}, \exp)$, that is, definable using restricted analytic functions and the exponential function. Denote by x the germ of the identity function. Then LE is defined to be the smallest subfield of $H(\mathbb{R}_{\mathrm{an}}, \exp)$ which is real closed, exp- and log-closed and contains $\mathbb{R}(x)$. It is the field of the germs of all compositions of semialgebraic functions, exp and log (cf. Example 6.16). As already mentioned, we know from [D–M–M1] that $H(\mathbb{R}_{\mathrm{an}}, \exp)$ is a Hardy field and a model of $T_{\mathrm{an}}(\exp)$. So we can take $M = H(\mathbb{R}_{\mathrm{an}}, \exp)$. With this choice, $H(\mathbb{R}_{\mathrm{an}}, \exp)$ is equal to $LE_{\mathcal{F}_{\mathrm{an}}}(x)$ (cf. Section 5 of [D–M–M1]), and the Hardy field LE is equal to $LE_{\mathcal{F}_{LE}}(x)$ (cf. Section 3 of [D–M–M2]). In this section and the next, we derive a positive solution to **Hardy's Conjecture** (cf. Theorem 6.49) from Theorem 6.44, and prove the undefinability of the Riemann ζ-function, working directly in $H(\mathbb{R}_{\mathrm{an}}, \exp)$.

Note that we do *not* define $LE_{\mathcal{F}}(x)$ to be the definable closure of $\mathbb{R}(x)$ inside of M. We know from [D–M–M1] that it will coincide with the definable closure in the case of $\mathcal{F} = \mathcal{F}_{\mathrm{an}}$. But in general, it will be properly contained in the definable closure. Indeed, the compositional inverse of the germ

$$\log(x) \log_2(x)$$

appearing in the Hardy problem is definable over LE, but not an LE-function. In fact it is *not even asymptotic* to an LE-function. This was conjectured by Hardy [HD2]. Denote by $i(x)$ the compositional inverse of $x \log(x)$. Then $\exp(i(x))$ is the compositional inverse of $\log(x) \log_2(x)$. Hardy's conjecture was first established for the function $\exp(\exp(i(x)))$ (instead of $\exp(i(x))$) in [S]. Then it was proved in [D–M–M2] for $\exp(i(x))$. This was done in the following main steps.

1) Identifying $i(x)$ with its germ, we have that $i(x) \in H(\mathbb{R}_{\mathrm{an}}, \exp)$. By an argument (using a theorem of Rosenlicht [RO4]) about Liouville extensions of the Hardy field $\mathbb{R}(x)$, Corollary 4.6 of [D–M–M2] shows that $i(x) \notin LE$.

2) Further, it is shown that there is a convergent power series $f(X,Y)$ (that is, $f(X,Y) \in \mathcal{F}_{\mathrm{an}}$) such that

$$i(x) = \frac{x}{\log(x)} \left(1 + f\left(\frac{\log_2(x)}{\log(x)}, \frac{1}{\log(x)}\right)\right).$$

This is established by the Implicit Function Theorem.

3) A large model of $T_{\mathrm{an}}(\exp)$, called the logarithmic-exponential power series field $\mathbb{R}((t))^{LE}$, is constructed ([D–M–M2], Section 2).

4) A truncation closed embedding ι of $H(\mathbb{R}_{\mathrm{an}}, \exp)$ and LE in $\mathbb{R}((t))^{LE}$ is constructed ([D–M–M2], Section 3).

5) Finally, it is shown that if $\exp(i(x))$ is asymptotic to an LE-function, then $\iota(i(x))$ must be a truncation of some $h \in \iota(LE)$, thus $\iota(i(x)) \in \iota(LE)$, so $i(x) \in LE$, a contradiction ([D–M–M2], Section 4).

The proof that we present now replaces the last three arguments by an argument concerning the residue fields of $H(\mathbb{R}_{\mathrm{an}}, \exp)$.

In the sequel, we will not distinguish the variable x from the germ x of the identity function. Note that if $f(X_1, \ldots, X_n)$ is definable in $(\mathbb{R}_{\mathrm{an}}, \exp)$ and $g_1, \ldots, g_n \in H(\mathbb{R}_{\mathrm{an}}, \exp)$ are the germs of the functions $g_1(x), \ldots, g_n(x)$, then the element $f(g_1, \ldots, g_n) \in H(\mathbb{R}_{\mathrm{an}}, \exp)$ is defined to be the germ of the function

$$x \mapsto f(g_1(x), \ldots, g_n(x)).$$

In this way, f is made into a function on $H(\mathbb{R}_{\mathrm{an}}, \exp)$. In particular, the element $f(x) \in H(\mathbb{R}_{\mathrm{an}}, \exp)$ is the germ of the function $f(x)$.

Before we prove Theorem 6.49, we observe the following general fact: Let $f, g : \mathbb{R} \to \mathbb{R}$ be definable in $(\mathbb{R}_{\mathrm{an}}, \exp)$. Assume that $\exp f(x)$ is asymptotic to $g(x)$. This is equivalent to $\lim_{x \to \infty} f(x) - h(x) = 0$, where $h : (r, \infty) \to \mathbb{R}$ for suitable $r \in \mathbb{R}$ is the function $\log g(x)$, which again is definable in $(\mathbb{R}_{\mathrm{an}}, \exp)$. This means that the function $f(x) - h(x)$ is ultimately smaller than every non-zero constant function. Equivalently, its germ $f - h$ in $H(\mathbb{R}_{\mathrm{an}}, \exp)$ is infinitesimal, or in other words, $v(f - h) > 0$.

Theorem 6.49 *The compositional inverse of* $\log(x) \log_2(x)$ *is not asymptotic to an LE-function.*

Proof Assume that $\exp i(x)$ is asymptotic to some $g(x) \in LE$. Then also $h(x) := \log g(x) \in LE$, and
$$v(i(x) - h(x)) > 0 .$$
But then,
$$v\left(\frac{i(x)}{x} - \frac{h(x)}{x}\right) > -v(x) > 0 .$$

Now let w be the convex valuation associated to the largest convex subgroup of $v(H(\mathbb{R}_{\mathrm{an}}, \exp))$ not containing $v(x)$. This contains $v(\log(x))$. Thus, $w(\log(x)) = 0$, and $w(x) < 0$. By convexity of w we have that
$$w\left(\frac{i(x)}{x} - \frac{h(x)}{x}\right) \geq -w(x) > 0 \qquad (74)$$
with $\frac{h(x)}{x} \in LE$. Now observe by 2) above that,
$$\frac{i(x)}{x} = \frac{1}{\log(x)}\left(1 + f\left(\frac{\log_2(x)}{\log(x)}, \frac{1}{\log(x)}\right)\right)$$
with $f \in \mathcal{F}_{\mathrm{an}}$. Thus,
$$\frac{i(x)}{x} \in \mathbb{R}(\log(x), \log_2(x))^{r\mathcal{F}_{\mathrm{an}}} \subset LE^w_{\mathcal{F}_{\mathrm{an}}}(\log(x)) .$$
By Corollary 6.47,
$$LE_{\mathcal{F}_{LE}}(x)w = LE^w_{\mathcal{F}_{LE}}(\log(x)) \subset LE^w_{\mathcal{F}_{\mathrm{an}}}(\log(x)) = LE_{\mathcal{F}_{\mathrm{an}}}(x)w .$$
It follows by (74) and Lemma 6.2 that
$$\frac{i(x)}{x} \in LE^w_{\mathcal{F}_{LE}}(\log(x)) \subset LE .$$
But then $i(x) \in LE$, a contradiction. This proves that $\exp i(x)$ is not asymptotic to any function in LE. $\square$

10 Undefinability of the Riemann ζ-function

Recall that the restriction to $(1, +\infty)$ of the **Riemann ζ-function** is given by:
$$\zeta(x) = \sum_{n=1}^{\infty} \frac{1}{n^x} = \sum_{n=1}^{\infty} \exp(-x \log n) .$$
It follows that there are positive elements $r_m \in \mathbb{R}$ such that for every $m > 1$,
$$\left|\zeta(x) - \sum_{n=1}^{m-1} \exp(-x \log n)\right| < r_m \exp(-x \log m)$$
holds for all large enough x.

In [D–M–M2] it is shown that the restriction of the ζ-function to $(1, +\infty)$ is not definable in $(\mathbb{R}_{\mathrm{an}}, \exp)$. The proof uses the truncation closed embedding of the Hardy field $H(\mathbb{R}_{\mathrm{an}}, \exp)$ in the logarithmic-exponential power series field. In this section, we wish to present an alternative proof, using our structure theorem for $H(\mathbb{R}_{\mathrm{an}}, \exp)$ instead.

We first make the following observation. If x is a positive element in the real closed field K, then it has a unique positive k-th root, for every $k \in \mathbb{N}$. So if K contains the real closure of a field $\mathbb{R}(x_i \mid i \in I)$, with all x_i positive, then $x_i^q \in K$ for all $i \in I$ and all $q \in \mathbb{Q}$. This can be used to show that every real closed field K, with its natural (or any convex) valuation v, admits a cross-section (that is, an embedding s of the group $v(K)$ in the multiplicative group $K^\times$ such that $v(s(\alpha)) = \alpha$ for all $\alpha \in v(K)$; cf. Chapter 4). Indeed, take any maximal set $\mathcal{X} = \{x_i \mid i \in I\} \subset K$ such that the values $v(x_i)$ are rationally independent. By the maximality of the set, together with Lemma 6.10, it follows that $v(K)$ is the divisible hull of $v(\mathbb{R}(x_i \mid i \in I)) = \bigoplus_{i \in I} \mathbb{Z} v(x_i)$. For every $\alpha \in v(\mathbb{R}(x_i \mid i \in I))$ there is a unique element x of the multiplicative group $\langle \mathcal{X} \rangle$ generated by the x_i, such that $v(x) = \alpha$. Consequently, there is a unique cross-section s of (K, v) whose image contains $\mathcal{X}$, and this image $s(v(K))$ is the divisible hull

$$\widetilde{\langle \mathcal{X} \rangle} = \left\{ \prod_{i \in I_0} x_i^{q_i} \mid I_0 \subset I \text{ finite}, q_i \in \mathbb{Q} \right\}$$

of $\langle \mathcal{X} \rangle$. If we have fixed a cross-section s, or a set $\mathcal{X}$ and take s to be the associated cross-section, then we call $\mathbb{R}^\times \cdot s(v(K))$ the set of **monomials** of K. Hence the monomials are the elements of the form

$$d = r \prod_{i \in I_0} x_i^{q_i} \text{ with } 0 \neq r \in \mathbb{R},\ I_0 \subset I \text{ finite, and } q_i \in \mathbb{Q} \text{ for every } i \in I_0.$$

We choose a representation $H(\mathbb{R}_{\mathrm{an}}, \exp) = \mathbb{R}(x_i \mid i \in I)^{\mathrm{r}\mathcal{F}_{\mathrm{an}}}$ with $v(x_i)$, $i \in I$, rationally independent, which exists by Theorem 6.32. We may assume in addition that x is among the x_i. We let s be the unique cross-section of $H(\mathbb{R}_{\mathrm{an}}, \exp)$ whose image contains $\{x_i \mid i \in I\}$; accordingly, we obtain the set of monomials.

Lemma 6.50 *Suppose that the element $h \in H(\mathbb{R}_{\mathrm{an}}, \exp)$ satisfies*

$$\left| h - \sum_{n=1}^{m-1} d_n \right| < r_m d_m \quad \text{for all } m > 1, \tag{75}$$

where $0 < r_n \in \mathbb{R}$, and the d_n are positive monomials such that the values $v(d_n)$ are strictly increasing. Then these values are contained in a finitely generated subgroup of $v(H(\mathbb{R}_{\mathrm{an}}, \exp))$.

Proof From Lemma 6.41 we infer that $h \in \mathbb{R}(x_i^{1/k} \mid i \in I_0)^{\mathrm{h}\mathcal{F}_{\mathrm{an}}} =: K$ for some $k \in \mathbb{N}$ and some finite subset $I_0 \subset I$, and that the value group of this field is the finitely generated subgroup $v(K) = \bigoplus_{i \in I_0} \mathbb{Z} \frac{v(x_i)}{k}$ of $v(H(\mathbb{R}_{\mathrm{an}}, \exp))$. From the rational independence of the values $v(x_i)$ it follows for every monomial d that $v(d) \in v(K)$ if and only if $d \in K$.

We write $S_0 = 0$ and

$$S_m := \sum_{n=1}^{m} d_n$$

for all $m \geq 1$. From (75) and $r_m \neq 0$ it follows that

$$v(h - S_{m-1}) \geq v(r_m d_m) = v(d_m).$$

for all $m > 1$. Hence,

$$v(h - S_{m-1} - d_m) = v(h - S_m) \geq v(d_{m+1}) > v(d_m),$$

which implies that

$$v(h - S_{m-1}) = v(d_m)$$

for all $m \geq 1$.

10. Undefinability of the Riemann ζ-function

Suppose that $v(d_n) \notin v(K)$ for some $n \in \mathbb{N}$, and take n to be the smallest integer with this property. Then $d_j \in K$ for $1 \leq j < n$. Consequently, $h - S_{n-1} \in K$. But $v(h - S_{n-1}) = v(d_n) \notin v(K)$, a contradiction. $\square$

For the application to the Riemann ζ-function, we run our construction of $LE_{\mathcal{F}_{an}}(x)$ with a slight refinement. We choose a $\mathbb{Q}$-basis $\mathcal{B}$ of $\mathbb{R}$ containing the linearly independent elements $\log(p)$, where p runs through all primes. Starting our construction from $K_0^u = \mathbb{R}(\log_m(x) \mid m \geq 0)^{r\mathcal{F}_{an}}$, we may first adjoin all elements $\exp(rx)$ as new x_i's. Indeed, as the elements $rx, \log_m(x), r \in \mathcal{B}, m \geq 1$, are rationally independent over the valuation ring, Lemma 6.12 shows that the values $v(\exp(rx))$, $v(\log_m(x))$, $r \in \mathcal{B}$, $m \geq 0$, are rationally independent. Hence, the values $v(\exp(rx))$, $r \in \mathcal{B}$, are rationally independent over $v(K_0^u)$. Therefore, for all $b \in \mathcal{B}$, $\exp(bx) \notin K_0^u(\exp(rx) \mid r \in \mathcal{B} \setminus \{b\})$. So we can assume the elements $\exp(x \log(p))$ to be among the x_i.

For $n \in \mathbb{N}$, we write $n = \prod_{p \text{ prime}} p^{\nu_p}$ with integers $\nu_p \geq 0$. Then we obtain that

$$d_n := \exp(-x \log(n)) = \exp\left(-x \log\left(\prod_{p \text{ prime}} p^{\nu_p}\right)\right)$$

$$= \exp\left(-x \sum_{p \text{ prime}} \nu_p \log(p)\right) = \prod_{p \text{ prime}} (\exp(x \log(p)))^{-\nu_p},$$

is a monomial. As the sequence $\log(n)$, $n \in \mathbb{N}$, is strictly increasing, Lemma 6.12 shows that also $v(d_n)$, $n \in \mathbb{N}$, is strictly increasing.

If ζ were definable in $(\mathbb{R}_{an}, \exp)$, it would follow from the foregoing lemma that the values $v(\exp(-x \log(n)))$ lie in a finitely generated group. But this is not true, since the values $v(\exp(-x \log(p))) = -v(\exp(x \log(p)))$, p prime, are rationally independent. This proves that the restriction of the ζ-function to $(1, +\infty)$ is not definable in $(\mathbb{R}_{an}, \exp)$.

Remark 6.51 In [D–S1], an o-minimal exponentially bounded expansion of the reals $(\mathbb{R}_{an*}, \exp)$ is given. In this expansion, the restriction of the ζ-function to $(1, +\infty)$ *is* definable.

APPENDIX A

The model theory of contraction groups

In this Appendix, the theory of abelian groups with contractions is axiomatized. By a detailed study of the algebraic properties of these groups, this theory is shown to be model complete, complete, decidable and to admit quantifier elimination. Further, the theory is shown to be weakly o-minimal, and yet *not* to have the algebraic exchange property.

For the discussion how to obtain contractions from exponentials, see Section 2.7. In this Appendix, we will take over the axiom system that was derived there to study the model theoretic properties of divisible ordered abelian groups with centripetal contractions. The property "centripetal" corresponds to the growth axiom scheme (GA) satisfied by the usual exponential. Note that here, for the sake of symmetry, we shall extend the domain of the contraction to be the whole group, whereas in Section 2.7 the domain of the contraction was just the negative cone of the group.

1 Preliminaries

In this appendix, we will assume $\mathbb{N}$**, the set of natural numbers, to contain** 0**. Further, whenever we will talk of ordered groups, we will mean ordered abelian groups.** We will omit the use of parenthesis to relax the notation. For example, we shall write $v_G g$ for $v_G(g)$ and χg for $\chi(g)$. We summarize some preliminary facts that we shall need.

Recall that the map v_G denotes the natural valuation on an ordered abelian group; it satisfies

(V0) $v_G x = \infty \iff x = 0$,

(V1) $v_G(x - y) \geq \min\{v_G x, v_G y\}$.

From these rules, we may deduce

(V2) $v_G x = v_G(-x)$,

(V3) $v_G\left(\sum_{1 \leq i \leq n} x_i\right) = \min_{1 \leq i \leq n} v_G x_i$ if all non-zero x_i have different values,

(V4) $v_G(x - y) > \min\{v_G x, v_G y\} \implies v_G x = v_G y$.

Note that for every $a \in G$ and every $n \in \mathbb{Z} \setminus \{0\}$, the element $na \in G$ is archimedean equivalent to a and so, the natural valuation satisfies the axiom scheme

(NV1) $v_G(nx) = v_G x \qquad (0 \neq n \in \mathbb{Z})$.

Recall that v_G is compatible with the order:

(NV2) $(v_G x < v_G y \Rightarrow |x| > |y|) \land (|x| \geq |y| \Rightarrow v_G x \leq v_G y)$.

Let us also note:

(NV3) $\text{sign}\left(\sum_{1 \leq i \leq n} x_i\right) = \text{sign}(x_m)$ if $v_G x_m < v_G x_i$ for all $i \neq m$,

(NV4) $v_G(x - y) > v_G x \implies \text{sign}(x) = \text{sign}(y)$.

Further, (NV3) may be generalized to

(NV5) $x_m < x'_m \implies \sum_{1 \leq i \leq n} x_i < \sum_{1 \leq i \leq n} x'_i$ if $v_G(x_m - x'_m) < v_G x_i$ and $v_G(x_m - x'_m) < v_G x'_i$ for all $i \neq m$.

2 Cuts in ordered Abelian groups

Let $(S, <)$ be a totally ordered set. If $S_1, S_2 \subset S$ and $a \in S$, we will write $a < S_2$ if $a < b$ for all $b \in S_2$, and further, we will write $S_1 < S_2$ if $a < S_2$ for all $a \in S_1$. Similarly, we use the relations $>$, $\leq$ and $\geq$. A pair (S_1, S_2) of two convex subsets of S satisfying $S_1 \cup S_2 = S$ will be called a **quasicut** in S if $S_1 \leq S_2$, and it will be called a **cut** if it even satisfies $S_1 < S_2$. We allow S_1 or S_2 to be empty, with the convention that $\emptyset < G$ and $G < \emptyset$. Suppose that an ordered set T contains S and that $b \in T$. We will say that b **realizes** (S_1, S_2) if $S_1 \leq b \leq S_2$ (in T). Further, $(\{a \in S \mid a \leq b\}, \{a \in S \mid a > b\})$ will be called the **cut induced by** b in S. Note that a (quasi)cut in a densely ordered set can be realized by at most one element of that set. In contrast to this, a cut in a discretely ordered set is either not realized in that set or is realized by exactly two elements of the set, the bigger one being the successor of the smaller.

If (G_1, G_2) is a quasicut in G and $g \in G$, then
$$(G_1, G_2) - g := (\{a - g \mid a \in G_1\}, \{a - g \mid a \in G_2\})$$
is again a quasicut in G, called a **shift** of (G_1, G_2). If (G_1, G_2) is a cut, then also $(G_1, G_2) - g$ is a cut. If (G_1, G_2) is the cut induced by b in G, then $(G_1, G_2) - g$ is the cut induced by $b - g$ in G. Note that two elements $a \neq a'$ may determine the same shift $(G_1, G_2) - a = (G_1, G_2) - a'$.

To every quasicut (G_1, G_2) in G, we may associate a quasicut in $v_G G$ in the following way. If G_2 contains an element ≤ 0, then we set $S_1 = v_G(G_1)$ and $S_2 = v_G(G^{<0} \cap G_2)$ and find that (S_1, S_2) is a quasicut in $v_G G$ by virtue of (NV2). If on the other hand, G_2 is contained in $G^{>0}$, then $-G_1$ contains an element ≤ 0, and we set $S_1 = v_G(-G_2) = v_G(G_2)$ and $S_2 = v_G(G^{<0} \cap -G_1) = v_G(G^{>0} \cap G_1)$; again, (S_1, S_2) is a quasicut in $v_G G$. If (S_1, S_2) is a cut, then (G_1, G_2) will be called a v_G-**cut**. From this definition, we may deduce the following criterion: (G_1, G_2) is a v_G-cut if and only if there is no pair (g_1, g_2) of elements $g_1 \in G_1$ and $g_2 \in G_2$ such that $v_G g_1 = v_G g_2$ and $\text{sign}(g_1) = \text{sign}(g_2)$. Using this criterion together with (NV2), we see that every element b in an extension of $(G, <)$ with $v_G b \notin v_G G \cup \{\infty\}$ will induce a v_G-cut in G. There is also a converse, stated as part b) of the following lemma:

Lemma A.1 *Let G be divisible and (G_1, G_2) be a cut in G.*

a) Among all shifts of (G_1, G_2), there is at most one v_G-cut. If (G_1, G_2) is realized by some element $a \in G$, then $(G_1, G_2) - a$ is this unique v_G-cut, realized by 0.

b) Assume (G_1, G_2) to be a v_G-cut which is not realized in G. If $(H, <)$ is an extension of $(G, <)$ and $b \in H$ realizes (G_1, G_2), then $v_G b \notin v_G G \cup \{\infty\}$.

Proof a): Suppose that $(G_1, G_2) - g$ and $(G_1, G_2) - g'$ are v_G-cuts; we have to show that they are equal. Replacing G_i by $G_i - g' = \{h - g' \mid h \in G_i\}$, $i = 1, 2$, and g by $g - g'$, we may assume from the start that $g' = 0$. Further, let us assume that $-g \in G_1$; for $-g \in G_2$, the proof is symmetrical.

Suppose that $(G_1 - g) \cap G_2 \neq \emptyset$ and choose $a \in G_1$ such that $a - g \in G_2$. If $a \geq -g$, it follows that $2a \in G_2$ since $2a \geq a - g$. If $a \leq -g$, it follows that $-2g \in G_2$ since $-2g \geq a - g$. In view of $a, -g \in G_1$, both are impossible by our criterion for v_G-cuts.

2. Cuts in ordered Abelian groups

Now suppose that $G_1 \cap (G_2 - g) \neq \emptyset$ and choose $b \in G_2$ such that $b - g \in G_1$. Then $b - 2g = (b - g) - g \in G_1 - g$. Applying our criterion to (G_1, G_2), we find that also $b/2 \in G_2$ and thus, $b/2 - g \in G_2 - g$. Applying our criterion to $(G_1, G_2) - g$, we obtain $b - 2g = 2(b/2 - g) \in G_2 - g$, contradicting $b - 2g \in G_1 - g$.

We have now shown that (G_1, G_2) and $(G_1, G_2) - g$ are equal. As to the last assertion of a), we leave it to the reader to show that $(G_1, G_2) - a$ is a v_G-cut realized by 0. The uniqueness then follows from what we have already proved.

b): Assume that b realizes (G_1, G_2) and that there is some $a \in G$ such that $v_G a = v_G b$, that is, a and b are archimedean equivalent. We may assume that $\text{sign}(a) = \text{sign}(b)$. Since G is assumed to be divisible, this implies the existence of some $q \in \mathbb{Q}^{>0}$ such that b lies between a and $qa \in G$. Since a and qa are archimedean equivalent, we have $v_G a = v_G qa$ and our above criterion shows that (G_1, G_2) cannot be a v_G-cut. □

Note that the proof shows that instead of requiring G to be divisible, it suffices to suppose that G is p-divisible for some prime p (or that all archimedean components are densely ordered). Without this condition, the lemma is not even true for densely ordered groups. Indeed, in the lexicographic product $\mathbb{Z} \amalg \mathbb{Q}$, both

$$(\{(x,y) \mid x \leq 0\}, \{(x,y) \mid x \geq 1\})$$

and

$$(\{(x,y) \mid x \leq -1\}, \{(x,y) \mid x \geq 0\}) = (\{(x,y) \mid x \leq 0\}, \{(x,y) \mid x \geq 1\}) - (1,0)$$

are v_G-cuts.

The cut (G_1, G_2) will be called a **shifted v_G-cut** if there exists some $g \in G$ such that $(G_1, G_2) - g$ is a v_G-cut. We have:

Corollary A.2 *Let the situation be as in the previous lemma. Assume that (G_1, G_2) is not realized in G but is realized by $b \in H$, where $(H, <)$ is an extension of $(G, <)$. Then (G_1, G_2) is a shifted v_G-cut if and only if there is some $g \in G$ such that $v_G(b - g) \notin v_G G \cup \{\infty\}$. In this case, $(G_1, G_2) - g$ is the unique v_G-cut among all shifts of (G_1, G_2).*

Proof Assume that $(G_1, G_2) - g$ is a v_G-cut. Since b realizes (G_1, G_2), the element $b - g$ realizes $(G_1, G_2) - g$. Hence by part b) of Lemma A.1, $v_G(b - g) \notin v_G G \cup \{\infty\}$. Conversely, if for some $g \in G$ we have $v_G(b - g) \notin v_G G \cup \{\infty\}$, then by our remark preceding Lemma A.1, $(G_1, G_2) - g$ is a v_G-cut. The uniqueness assertion follows directly from Lemma A.1. □

The model theoretical facts that we will use in this appendix, can be found in most of the books on general model theory, e.g., [C–K] or [PR2]. Also, they can be found in [KF3]. In particular, let us mention the following facts (cf. Korollar 2.19 of [PR2], or [KF3]):

Lemma A.3 *Let $\mathcal{S}'$ be a substructure of the structure $\mathcal{S}$. If $\mathcal{S}'$ is existentially closed in $\mathcal{S}$, then $\mathcal{S}$ is embeddable over $\mathcal{S}'$ in every $|\mathcal{S}|^+$-saturated elementary extension of $\mathcal{S}'$. Conversely, if $\mathcal{S}$ is embeddable over $\mathcal{S}'$ in some elementary extension of $\mathcal{S}'$, then $\mathcal{S}'$ is existentially closed in $\mathcal{S}$.*

Note that $|\mathcal{S}|^+$ denotes the successor cardinal of the cardinality of (the underlying set of) $\mathcal{S}$.

We will need the following well-known facts about ordered groups. Their proofs can be found in [KF3].

Lemma A.4 *Let $(G,<) \subset (H,<)$ be an extension of ordered abelian groups and $b \in H \setminus G$. Assume that G be divisible. If $b' \notin G$ is also an element of some extension of $(G,<)$, inducing in G the same cut as b, then the assignment $b \mapsto b'$ defines an order preserving isomorphism from $G + \mathbb{Z}b$ onto $G + \mathbb{Z}b'$. On the other hand, in every $|G|^+$-saturated extension of $(G,<)$ there is an element $b' \notin G$ inducing in G the same cut as b.*

(The assertions of this lemma can easily be used to show the model completeness of the elementary class of non-trivial divisible ordered abelian groups.)

Lemma A.5 *Let $(G,<) \subset (H,<)$ and $(G,<) \subset (H',<)$ be extensions of ordered abelian groups. Suppose that b_i and b'_i, $i \in I$, are elements of H and H' respectively, such that*
1) *all $v_G b_i$, $i \in I$, are different and not contained in $v_G G$,*
2) *the assignment $v_G b_i \mapsto v_G b'_i$ extends the identity of $v_G G$ to an order isomorphism*
$$\tau : v_G G \cup \{v_G b_i \mid i \in I\} \to v_G G \cup \{v_G b'_i \mid i \in I\}\,,$$
3) *$\text{sign}(b_i) = \text{sign}(b'_i)$ for all $i \in I$.*

Then the assignment $b_i \mapsto b'_i$ defines order and valuation preserving isomorphisms

$$G + \sum_{i \in I} \mathbb{Z}b_i \longrightarrow G + \sum_{i \in I} \mathbb{Z}b'_i$$

$$G + \sum_{i \in I} \mathbb{Q}b_i \longrightarrow G + \sum_{i \in I} \mathbb{Q}b'_i$$

both inducing τ. (Here, the groups are endowed with the restrictions of the orders of the divisible hulls of H and H' which are uniquely determined by those of H and H'.) Note that $v_G G \cup \{v_G b_i \mid i \in I\}$ is the value set of both groups on the left-hand side, and $v_G G \cup \{v_G b'_i \mid i \in I\}$ is the value set of both groups on the right-hand side.

Lemma A.6 *Let $(H,<)$ be generated by n elements over $(G,<)$. Then the set $v_G H \setminus v_G G$ consists of at most n values.*

3 Ordered abelian groups with contractions

In this section, the axioms for contractions on ordered abelian groups will be listed and discussed. Afterwards, we study precontraction groups (on which the contraction is not assumed to be surjective) and their extensions. By means of embedding lemmas, the model theory of non-trivial divisible centripetal contraction groups will be shown to be model complete, complete, decidable and to admit quantifier elimination (Theorems A.29, A.32, A.33 and A.30).

We will work in the language $\mathcal{L}_{\text{cg}} = \{+, -, 0, <, \chi\}$ where $<$ is a binary relation symbol, $+$ is a binary function symbol, and $-$ and χ are unary function symbols. If $(G, +, -, 0, <, \chi)$ is an $\mathcal{L}_{\text{cg}}$-structure, then it will be called a **precontraction group** if it satisfies

(OAG) $(G, +, -, 0, <)$ is an ordered abelian group,

(C0) $\chi x = 0 \iff x = 0$,

(C$\leq$) χ preserves $\leq$,

(C$-$) $\chi(-x) = -\chi x$,

(CA) if x is archimedean equivalent to y and $\operatorname{sign}(x) = \operatorname{sign}(y)$, then $\chi x = \chi y$.

If these axioms hold, then χ will be called a **precontraction**. If in addition,

(CS) χ is surjective,

then χ will be called a **contraction** and the group is a **contraction group**. Axioms (CA) and (CS) together show that every archimedean ordered contraction group must be trivial. Further, $(G, +, -, 0, <, \chi)$ will be called **centripetal** if it satisfies

(CP) $x \neq 0 \implies |x| > |\chi x|$,

and it will be called **centrifugal** if it satisfies

(CF) $x \neq 0 \implies |x| < |\chi x|$.

In what follows, we will write (G, χ) instead of $(G, +, -, 0, <, \chi)$.

In the presence of axiom (C−), axiom (CA) may be expressed by the following recursive elementary axiom scheme:

$$x \geq y > 0 \wedge ny \geq x \implies \chi x = \chi y \qquad (n \in \mathbb{N}).$$

Observe that axioms (C0) and (C≤) imply

(CSN) $\operatorname{sign}(\chi x) = \operatorname{sign}(x)$

and that axioms (C0), (C−) and (CA) together imply

(CZ) $\chi(zx) = \operatorname{sign}(z) \cdot \chi x \qquad (z \in \mathbb{Z})$.

Observe further that by axiom (C≤),

(CC) $(x \leq y \leq z \wedge \chi x = \chi z) \implies \chi y = \chi z$

and more generally, that the preimage of every convex set under χ is convex. This shows that in the presence of axiom (C≤), the axiom scheme (CA) may be replaced by the single axiom

(CA′) $\chi(2x) = \chi x$.

This and all other axioms are immediately seen to be elementary in the language $\mathcal{L}_{\mathrm{cg}}$. All axioms are universal, except for the surjectivity axiom (CS). Since properties described by universal axioms are inherited by substructures, we have:

Lemma A.7 *Every substructure S of a precontraction group (G, χ) is again a precontraction group. If (G, χ) is centripetal (resp. centrifugal), then so is S.*

We will need a further axiom scheme which is not universal. Namely, that $(G, +, 0)$ is divisible:

(D) $\forall x \, \exists y : ny = x \qquad (0 \neq n \in \mathbb{N})$.

We will consider the theory of divisible centripetal contraction groups. Some results also hold for the theory of divisible centrifugal contraction groups. We will frequently need that the groups are non-trivial, that is, they satisfy the axiom $\exists x : x \neq 0$. Before we continue, let us put together some technical preliminaries. Using the natural valuation v_G, we may express axiom (CA) in the following way:

(CV1) $(v_G x = v_G y \wedge \operatorname{sign}(x) = \operatorname{sign}(y)) \implies \chi x = \chi y$

(but note that v_G is neither a symbol in our language $\mathcal{L}_{\mathrm{cg}}$ nor definable in the theory of divisible centripetal or centrifugal contraction groups). Sometimes, we will only be interested in equality up to the sign; instead of writing $|a| = |b|$ we will then write $a = \pm b$. Then, (CV1) reads as follows:

(CV2) $v_G x = v_G y \implies \chi x = \pm \chi y$.

From (V3), (NV3) and (V4), (NV4) we may infer:

(CV3) $\chi\left(\sum_{1\leq i\leq n} x_i\right) = \chi x_m$ if $v_G x_m < v_G x_i$ for all $i \neq m$,

(CV4) $v_G(x-y) > v_G x \implies \chi x = \chi y$.

From (CV2) and (NV2) together with (C≤), one may deduce

(CV5) $v_G x \leq v_G y \implies |\chi x| \geq |\chi y|$.

The following lemma collects some generalities about precontractions.

Lemma A.8 *Let (G, χ) be a precontraction group. Then the following assertions hold:*

a) sign(a) = sign(χa) *for every $a \in G$, hence $\chi G^{<0} \subset G^{<0}$ and $\chi G^{>0} \subset G^{>0}$. Moreover, $\chi G^{<0} = -\chi G^{>0}$.*

b) χ is centripetal if and only if $v_G(\chi a) > v_G a$ for all $a \in G \setminus \{0\}$. Similarly, χ is centrifugal if and only if $v_G(\chi a) < v_G a$ for all $a \in G \setminus \{0\}$.

c) Every non-trivial centripetal precontraction group is densely ordered. The same is true for every non-trivial centrifugal contraction group.

Proof a): If $a < 0$ then by (C≤) and (C0), $\chi a < \chi 0 = 0$. Similarly, $a > 0$ is shown to imply $\chi a > 0$. The last assertion follows directly from (C−).

b): Assume that χ is centripetal. Then for every $a \in G \setminus \{0\}$, we have $|\chi a| < |a|$ and thus, $v_G(\chi a) \geq v_G a$. But for every $b \in G$ with $v_G a = v_G b$ it follows from (CA) and (CP) that $|\chi a| = |\chi b| < |b|$. Hence, $v_G(\chi a) = v_G a$ is impossible. The converse follows from (NV2). The proof for centrifugal precontractions is similar.

c): A centripetal precontraction group cannot have a least positive element 1 since by (C0), (C≤) and (CP), $0 < \chi g < g$ for every $g \in G^{>0}$. Now assume 1 to be the least positive element in a centrifugal contraction group (G, χ). Then there is some $a \in G$ such that $\chi a = 1$. But then by (C0), (C≤) and (CF), $0 < a < 1$, a contradiction. □

Lemma A.9 *Assume $(G, \chi) \subset (H, \chi)$ to be an extension of precontraction groups. Let $b \in H$ be such that $v_G b \notin v_G G \cup \{\infty\}$ and $\chi b = a \in G$. Then $(G+\mathbb{Z}b, \chi)$ is a precontraction group with $\chi(G+\mathbb{Z}b) = \chi G \cup \{a, -a\} \subset G$. Moreover, the extension of χ from (G, χ) to $G+\mathbb{Z}b$ is uniquely determined by the assignment $\chi b = a$.*

Proof Since $v_G b \notin v_G G$, (V3) and (NV1) show that every element in $G + \mathbb{Z}b$ has value either in $v_G G \cup \{\infty\}$ or equal to $v_G b$. Take any $d = g + nb \in G + \mathbb{Z}b$. Then $\chi d = \chi nb = \text{sign}(n) \cdot a$ if $v_G d = v_G b$, and $\chi d = \chi g$ if $v_G d = v_G g \in v_G G \cup \{\infty\}$. Hence, the extension of χ from G to $G + \mathbb{Z}b$ is uniquely determined by $\chi b = a$. It also proves $\chi(G + \mathbb{Z}b) = \chi G \cup \{a, -a\} \subset G$. Consequently, $G + \mathbb{Z}b$ is closed under χ and thus a precontraction group by virtue of Lemma A.7. □

Lemma A.10 *Assume $(G, \chi) \subset (H, \chi)$ to be an extension of precontraction groups. Let $a \in G \setminus \chi G$ and $b \in H$ be such that $a = \chi b$. Then $v_G b \notin v_G G \cup \{\infty\}$, and $(G + \mathbb{Z}b, \chi)$ is a precontraction group satisfying $\chi(G + \mathbb{Z}b) \subset G$. Moreover, if $(G, \chi) \subset (H', \chi')$ is a second extension of precontraction groups and $b' \in H'$ is such that $a = \chi' b'$, then the assignment $b \mapsto b'$ induces an isomorphism $G + \mathbb{Z}b \to G + \mathbb{Z}b'$ of precontraction groups over G.*

3. Ordered abelian groups with contractions

Proof If there would exist some $c \in G$ such that $v_G b = v_G c$ then in view of (CV2), $a = \chi b = \pm \chi c \in \chi G$. Hence, $a \notin \chi G$ implies $v_G b \notin v_G G \cup \{\infty\}$, and from Lemma A.9 it follows that $(G+\mathbb{Z}b, \chi)$ is a precontraction group with $\chi(G+\mathbb{Z}b) \subset G$.

We show that the ordering on $v_G G \cup \{v_G b\}$ is uniquely determined by (G, χ) and the assignment $\chi b = a$. Assume w.l.o.g. that $a \in G^{<0}$. Then $b \in H^{<0}$ by part a) of Lemma A.8. Let $c \in G^{<0}$. Since $v_G c \neq v_G b$, (NV2) shows that $v_G c < v_G b$ if and only if $c < b$ and $v_G c > v_G b$ if and only if $c > b$. Then by virtue of (C$\leq$), $v_G c < v_G b$ if and only if $\chi c < a$ and $v_G c > v_G b$ if and only if $\chi c > a$.

Now let b' be as in the hypothesis. By what we have proved already, $v_G b \notin v_G G \cup \{\infty\}$ and $v_G b' \notin v_G G \cup \{\infty\}$, and the assignment $v_G b \mapsto v_G b'$ induces an order isomorphism over $v_G G$ of the two sets $v_G G \cup \{v_G b\}$ and $v_G G \cup \{v_G b'\}$ (which are endowed with the restrictions of the ordering of $v_G H$ resp. $v_G H'$). By (CSN), $\text{sign}(b) = \text{sign}(a) = \text{sign}(b')$. From Lemma A.5 it now follows that the groups $G + \mathbb{Z}b$ and $G + \mathbb{Z}b'$, endowed with the restrictions of the ordering of H resp. H', are isomorphic over G by sending b to b'. By the uniqueness statement of Lemma A.9, this isomorphism is also an isomorphism of precontraction groups. $\square$

There are very simple examples for precontraction groups. The map "sign" is itself a precontraction on $\mathbb{Z}$, but it is neither centrifugal nor centripetal. It satisfies $\forall x \in G : |x| \geq |\chi x|$, but this does not remain true in extensions of $\mathbb{Z}$ where 1 is not the least positive element. $\mathbb{Z}$ does not admit a centrifugal or centripetal precontraction since it is archimedean. Indeed, it follows from part b) of Lemma A.8 that a non-trivial centrifugal or centripetal precontraction group must have an infinite value set.

Consider the Hahn sums $\coprod_\mathbb{N} \mathbb{Z}$ and $\coprod_{-\mathbb{N}} \mathbb{Z}$ where $\mathbb{Z}$ stands for the ordered group $(\mathbb{Z}, +, -, 0, <)$, and $\mathbb{N}$ resp. $-\mathbb{N}$ stands for the positive resp. negative integers with their usual ordering. On the first Hahn sum, we may define a precontraction in the following way: if $(a_i)_{i \in \mathbb{N}} \in \coprod_\mathbb{N} \mathbb{Z}$ and if i_0 is the minimal index such that $a_{i_0} \neq 0$, then we set

$$\chi(a_i)_{i \in \mathbb{N}} = \text{sign}(a_{i_0}) \cdot e_{i_0+1}$$

where e_i denotes the element of $\coprod_\mathbb{N} \mathbb{Z}$ which has a 1 at the index i and zeros everywhere else. The resulting centripetal precontraction group will be denoted by $\mathcal{P}_{\text{cp}}$. Analogously, if $(a_i)_{i \in -\mathbb{N}} \in \coprod_{-\mathbb{N}} \mathbb{Z}$ and if i_0 is the maximal index such that $a_{i_0} \neq 0$, then we set

$$\chi(a_i)_{i \in \mathbb{N}} = \text{sign}(a_{i_0}) \cdot e_{i_0-1}$$

where $e_i \in \coprod_{-\mathbb{N}} \mathbb{Z}$ is defined as above. The resulting centrifugal precontraction group will be denoted by $\mathcal{P}_{\text{cf}}$. These examples are as representative as they can be:

Lemma A.11 $\mathcal{P}_{\text{cp}}$ *is the prime structure of the elementary class of non-trivial centripetal precontraction groups. Analogously, $\mathcal{P}_{\text{cf}}$ is the prime structure of the elementary class of non-trivial centrifugal precontraction groups.*

Proof We will show that every substructure S of a centripetal precontraction group generated by one element $a > 0$ is isomorphic to $\mathcal{P}_{\text{cp}}$. Indeed, we infer from part b) of Lemma A.8 that $v_G(\chi^n a)$, $n \in \mathbb{N}$, is a strictly increasing sequence in the value set $v_G S$. Let S_0 be the subgroup of S which is generated by all the elements $\chi^n a$. Every element s of S_0 is then a finite sum $\sum_{j=0}^{\infty} \nu_j \chi^j a$ with coefficients $\nu_j \in \mathbb{Z}$, only finitely many of them non-zero. In view of (V3) and (NV1), we find

$v_G s = v_G(\chi^{j_0} a)$ if $j_0 = \min\{j \in \mathbb{N} \mid \nu_j \neq 0\}$. Then it follows from (CV2) that $\chi s = \pm \chi^{j_0+1} a$. This shows that S_0 is closed under χ and hence, $S_0 = S$.

By Lemma A.8 b), $v_G \chi^{n+1} a > v_G \chi^n a$ for all $n \in \mathbb{N}$. Moreover, by (CSN), $\text{sign}(\chi^n a) = \text{sign}(a) = 1 = \text{sign}(e_n)$ for all $n \in \mathbb{N}$. Hence by Lemma A.5, the assignment $\chi^n a \mapsto e_n$ extends linearly to an order and valuation preserving isomorphism from S onto $\coprod_{\mathbb{N}} \mathbb{Z}$.

The proof for the centrifugal case is similar. In this case, the sequence $v_G(\chi^n a)$, $n \in \mathbb{N}$, is strictly decreasing, and we have to set $j_0 = \max\{j \in \mathbb{N} \mid \nu_j \neq 0\}$ and send $\chi^n a$ to e_{-n}. □

An extension $(G, v_G) \subset (H, v_G)$ of valued groups is called **value set preserving** if $v_G H = v_G G$ (precisely speaking, if the induced embedding of $v_G G$ in $v_G H$ is onto). For example, the divisible hull $\tilde{G}$ of G is a value set preserving extension of (G, v_G) since every element in $\tilde{G}$ is archimedean equivalent to some element of G. But if $v_G a = v_G b$, then every precontraction on H will satisfy $\chi a = \pm \chi b$ by virtue of (CV2). On the other hand, if χ is a precontraction on G, then it may be extended to H by setting $\chi a = \text{sign}(a) \cdot \chi |b|$. This yields $\chi H = \chi G$, showing that H is closed under χ and thus, that (H, χ) is a precontraction group. Hence, we have:

Lemma A.12 *Let (G, χ) be a precontraction group. Then for every extension $(H, <)$ of $(G, <)$ which is value set preserving with respect to the natural valuation, χ extends in a unique way to a precontraction χ on H (under preservation of the properties "centripetal" and "centrifugal"), and we have $\chi H = \chi G$. More precisely, if $(G, \chi) \subset (H', \chi')$ is an extension of precontraction groups such that $(H, <) \subset (H', <)$, then the restriction of χ' to H coincides with χ.*

In particular, these assertions hold for $H = \tilde{G}$.

If the value set preserving extension $G \subset H$ is non-trivial, then $\chi H = \chi G \neq H$. Hence:

Corollary A.13 *Let $(G, \chi) \subset (H, \chi)$ be a proper value set preserving extension of precontraction groups. Then (H, χ) cannot be a contraction group. In particular, this holds for the divisible hull of a non-divisible precontraction group.*

To prove even more, we introduce a new map which is associated to χ. We define $\rho_\chi : v_G G \to G^{<0}$ as follows: if $\alpha = v_G a \in v_G G$, $a \in G^{<0}$, then $\rho_\chi \alpha = \chi a$. This is well-defined since by (CV1), the definition does not depend upon the choice of the negative element a of value α. We have $\rho_\chi(v_G G) = \chi G^{<0}$: if $b = \chi a \in \chi G$, then $b = \rho_\chi v_G a$. Hence, ρ_χ is surjective if and only if χ is. Moreover, ρ_χ preserves $\leq$: on the one hand, χ preserves $\leq$ by (C$\leq$); on the other hand, $v_G a < v_G a'$ for $a, a' \in G^{<0}$ implies $a < a'$.

Theorem A.14 *Let $(G, \chi) \subset (H, \chi)$ be a non-trivial extension of precontraction groups. If (H, χ) is a contraction group, then $v_G H \setminus v_G G$ is infinite and thus, $(G, v_G) \subset (H, v_G)$ is not value set preserving.*

Proof Since (H, χ) is a contraction group, we have $\chi H = H$. By assumption, $G \subset H$ is a non-trivial extension, hence there is an element $b \in H^{<0} \setminus G^{<0}$. Since $\rho_\chi(v_G H) = H^{<0}$, there is some $\beta \in v_G H \setminus v_G G$ such that $\rho_\chi \beta = b$. The fact that $v_G H \setminus v_G G$ is non-empty implies that $H^{<0} \setminus G^{<0}$ is infinite. Again by $\rho_\chi(v_G H) = H^{<0}$, it follows that $v_G H \setminus v_G G$ is infinite. □

3. Ordered abelian groups with contractions

We will now show that every precontraction group is embeddable in a divisible contraction group.

Lemma A.15 *Every (centripetal resp. centrifugal) precontraction group (G, χ) is embeddable in a divisible (centripetal resp. centrifugal) contraction group (H, χ).*

Proof (H, χ) will be the union over a chain of precontraction groups (G_n, χ), $n \in \mathbb{N}$. Let G_1 be the Hahn product $\mathbf{H}_{v_G G}\mathbb{R}$. Then there is an order and valuation preserving embedding of G in G_1, and the extension $(G, v_G) \subset (G_1, v_G)$ is value set preserving. Hence, there is a unique extension of χ from G to G_1.

Having constructed (G_n, χ), let us show how to obtain (G_{n+1}, χ). We set
$$\Gamma := v_G G_n \cup (G_n^{<0} \setminus \chi G_n^{<0}) \ .$$
To define an ordering on Γ, we extend the orderings which already exist on $v_G G_n$ and on G_n. Hence, we only have to give the order relation between two elements $\beta \in v_G G_n$ and $a \in G_n^{<0} \setminus \chi G_n^{<0}$. We set $a < \beta$ if $a < \rho_\chi \beta$; otherwise, we set $a > \beta$. Now, we let G_{n+1} be the Hahn product $\mathbf{H}_\Gamma \mathbb{R}$. Then there is a natural order and valuation preserving embedding of G_n in G_{n+1} which is induced by the embedding $v_G G_n \subset \Gamma$. We identify G_n with its image in G_{n+1}. It remains to define the extension of χ. Let $b \in G_{n+1}^{<0}$. If $v_G b = v_G b' \in v_G G_n$ for some $b' \in G_n^{<0}$ then necessarily, in view of (CV1), $\chi b = \chi b'$. If $v_G b = a \in G_n^{<0} \setminus \chi G_n^{<0}$, then we set $\chi b = a \in G_n \subset G_{n+1}$. For $b \in G_{n+1}^{>0}$, we set $\chi b = -\chi(-b)$. In this way, every element of G_n becomes an element of the range of χ in G_{n+1}. Consequently, χ will be surjective on the union of the G_n. Since all other axioms are universal, we see that this union is a contraction group. $\square$

Let us now consider the question whether there are "closures" of precontraction groups in contraction groups. Let $(G, \chi) \subset (G', \chi)$ be an extension of precontraction groups. We will call (G', χ) a **contraction hull** of (G, χ) if it is a contraction group and has the following universal property:

(CH) if $(G, \chi) \subset (H, \chi)$ is any extension of precontraction groups and (H, χ) is a contraction group, then there is an embedding of (G', χ) in (H, χ) over (G, χ).

Similarly, we will call (G', χ) a **divisible contraction hull** of (G, χ) if it is a divisible contraction group and has the following universal property:

(CHD) if $(G, \chi) \subset (H, \chi)$ is any extension of precontraction groups and (H, χ) is a divisible contraction group, then there is an embedding of (G', χ) in (H, χ) over (G, χ).

Note that if (G', χ) is a contraction hull of a divisible precontraction group (G, χ) but is not itself divisible, then its divisible hull is *not* a divisible contraction hull of (G, χ) (cf. Theorem A.14).

Lemma A.16 *For every precontraction group (G, χ) there exists a contraction hull and a divisible contraction hull. Such a hull (G', χ) may be chosen such that ρ_χ induces an order preserving bijection $v_G G' \setminus v_G G \to G'^{<0} \setminus \chi G$. Moreover, we can assume that for every $b \in G'$ there is some $n \in \mathbb{N}$ such that $\chi^n b \in G$.*

Proof By Lemma A.15, we may embed (G, χ) in some divisible contraction group (H, χ). We will construct a subgroup of H as follows. For the construction of the contraction hull, we set $G_0 = G$. For the construction of the divisible contraction hull, we let G_0 be the divisible hull of G and endow it with the unique extension of the precontraction χ (cf. Lemma A.12). Note that $\chi G_0 = \chi G \subset G$. Assume that G_ν for some ordinal ν is already constructed such that ρ_χ induces a bijection $v_G G_\nu \setminus v_G G \to \chi G_\nu^{<0} \setminus \chi G$ (for $\nu = 0$, this bijection is empty). If χ is not surjective on G_ν and hence there is some $a \in G_\nu^{<0} \setminus \chi G_\nu$, then we choose $b \in H$ such that $\chi b = a$, and we set $G_{\nu+1} := G_\nu + \mathbb{Z}b$ (resp. $G_{\nu+1} := G_\nu + \mathbb{Q}b$ if we are constructing the divisible contraction hull). From Lemma A.9 we know that $G_{\nu+1}$ together with the restriction of χ is a precontraction group having $v_G G_\nu \cup \{v_G b\}$ as its value set and satisfying $\chi G_{\nu+1} = \chi G_\nu \cup \{\pm a\} \subset G_\nu$. The map ρ_χ sends $v_G b$ to $\chi b = a$ which is the only element in $\chi G_{\nu+1}^{<0} \setminus \chi G_\nu = \{a\}$. Hence, ρ_χ induces a bijection $v_G G_{\nu+1} \setminus v_G G \to \chi G_{\nu+1}^{<0} \setminus \chi G$. Since $\chi G_{\nu+1} \subset G_\nu$, $G_{\nu+1}$ inherits the property that for every element b there is some $n \in \mathbb{N}$ such that $\chi^n b \in G$.

If λ is a limit ordinal and if we have constructed G_ν for all $\nu < \lambda$, then we let G_λ be the union over all G_ν. Then G_λ is again a precontraction group (the theory of precontraction groups being universal). Still, ρ_χ induces a bijection $v_G G_\lambda \setminus v_G G \to \chi G_\lambda^{<0} \setminus \chi G$, and for every $b \in G_\lambda$ there is some $n \in \mathbb{N}$ such that $\chi^n b \in G$. Since at every step we are constructing a non-trivial extension but remain in H, this process is bounded by the successor cardinal κ^+ of the cardinality κ of H. Hence, we will arrive at some G_μ for an ordinal $\mu < \kappa^+$ where χ is surjective. That is, (G_μ, χ) is a contraction group (resp. a divisible contraction group). We set $G' := G_\mu$. Since now $G' = \chi G'$ holds, we have that ρ_χ induces a bijection $v_G G' \setminus v_G G \to G'^{<0} \setminus \chi G$ (which is order preserving since ρ_χ preserves $\leq$).

Now let $(G, \chi) \subset (H', \chi')$ be any extension of precontraction groups and assume that (H', χ') is a contraction group (resp. a divisible contraction group). Assume that we have already embedded (G_ν, χ) in (H', χ') over (G, χ); we may identify it with its image in (H', χ'). Now we have to show how this embedding extends to $(G_{\nu+1}, \chi)$. Let $a \in G_\nu$ and $b \in H$ be chosen as above. Choose $b' \in H'$ such that $\chi' b' = a$. Then Lemma A.10 shows that the precontraction groups $(G_{\nu+1}, \chi)$ and $(G_\nu + \mathbb{Z}b', \chi')$ (resp. $(G_\nu + \mathbb{Q}b', \chi')$, using also Lemma A.12) are isomorphic over (G_ν, χ). For a limit ordinal λ, the embeddings of the (G_ν, χ), $\nu < \lambda$, extend canonically to an embedding of (G_λ, χ). It follows that (G', χ) is embeddable in (H', χ') over (G, χ). $\square$

Note that the universal property of the contraction hull is somewhat weak: the embedding is not necessarily unique. Indeed, we have seen in the proof above that there is some arbitrariness in the construction of the isomorphism, namely, the choice of $b' \in H'$ satisfying $\chi' b' = a$ is not canonical: we can take any $b'' \in H'$ with $v_G b'' = v_G b'$ instead. So we are not able to deduce the uniqueness of the contraction hull from the usual standard argument of category theory. However, uniqueness is not needed for our further results.

Applying the last lemma to the prime structures $\mathcal{P}_{cp}$ and $\mathcal{P}_{cf}$, we obtain:

Corollary A.17 *The elementary classes of non-trivial centripetal (resp. centrifugal) contraction groups and of non-trivial divisible centripetal (resp. centrifugal) contraction groups have prime models.*

Further, the contraction hulls have a good model theoretic property:

Lemma A.18 *Let $(G, \chi) \subset (G'', \chi)$ be an extension of precontraction groups and assume that (G, χ) is a contraction group (resp. a divisible contraction group). If (G, χ) is existentially closed in (G'', χ), then it is existentially closed in every contraction hull (resp. divisible contraction hull) (G', χ) of (G'', χ).*

Proof If (G, χ) is existentially closed in (G'', χ), then by Lemma A.3, (G'', χ) is embeddable in every $|G''|^+$-saturated elementary extension $(G, \chi)^*$ of (G, χ). We

3. Ordered abelian groups with contractions

identify (G'', χ) with its image in $(G, \chi)^*$. As an elementary extension, $(G, \chi)^*$ is a (divisible) contraction group like (G, χ). It follows from the universal property of (divisible) contraction hulls that every (divisible) contraction hull (G', χ) of (G'', χ) is embeddable in $(G, \chi)^*$ over (G'', χ). Again by Lemma A.3, it now follows that (G, χ) is existentially closed in (G', χ). □

Let $(G, \chi) \subset (H, \chi)$ be an extension of precontraction groups. For a given $b \in H$, we will consider the cut

$$(\{g \mid G \ni g \leq b\}, \{g \mid G \ni g > b\}) \tag{76}$$

induced by b in G and its image under χ:

$$(\{\chi g \mid G \ni g \leq b\}, \{\chi g \mid G \ni g > b\}) . \tag{77}$$

By virtue of (C≤),

$$\{\chi g \mid G \ni g \leq b\} \leq \chi b \leq \{\chi g \mid G \ni g > b\} . \tag{78}$$

Now assume (G, χ) to be a contraction group. Then $\chi G = G$ and (77) is a quasicut in G which is realized by χb. If the sets of (77) have a non-empty intersection, then it consists of just one element which must be χb, implying that $\chi b \in G$. If the intersection of the two sets is empty, then (77) is a cut. In particular, this is the case if $\chi b \notin G$ and then, (77) is the cut induced by χb in G. If (77) is a cut which is not realized by any element of G, then $\chi b \notin G$ and the cut (76) determines uniquely the cut induced by χb in G. But if there is an element $a \in G$ which realizes (77), then the cut (76) may not determine whether $\chi b \in G$. Indeed, we then only know that there is no element of G between a and χb. This happens if $a = \chi b$, but it may also happen that H contains an element d such that $v_G d > v_G G$ and $\chi b = a + d$. Let us discuss the situation in general:

Lemma A.19 *Let $(G, <) \subset (H, <)$ be an extension of ordered groups. Assume that two elements $b_1 \in H$ and $b_2 \in G$ realize the same quasicut in $(G, <)$. Then there is no element of G properly between b_1 and b_2; hence, there is no element of G properly between $b_1 - b_2$ and 0. We have two cases:*

a) If $b_1 - b_2$ is archimedean equivalent to some element of $G \setminus \{0\}$ (i.e., $v_G(b_1 - b_2) \in v_G G$), then $v_G(b_1 - b_2)$ is the maximal element in $v_G G$ and $(G, <)$ is discretely ordered.

b) If $b_1 - b_2$ is not archimedean equivalent to some element of $G \setminus \{0\}$, then it is archimedean smaller than all non-zero elements of G, that is, $v_G(b_1 - b_2) > v_G G$. If $b_2 \neq 0$, then it follows that $v_G b_1 = v_G b_2$ and $\text{sign}(b_1) = \text{sign}(b_2)$.

In particular, if $(G, <)$ is dense and $v_G G$ is cofinal in $v_G H$ then necessarily, $b_1 = b_2$.

Proof If there were an element $b' \in G$ properly between b_1 and b_2, then both b' and b_2 would realize the quasicut in G. This can only hold if G is discretely ordered, one of the two elements is the maximum of the lower cut set and the other is the minimum of the upper cut set. But then, b_1 must lie between b' and b_2, a contradiction. Hence, there is no element of G properly between b_1 and b_2, and there is no element of G properly between $b_1 - b_2$ and 0.

If there were $a \in G \setminus \{0\}$ such that $v_G a > v_G(b_1 - b_2)$ then a would be archimedean smaller than $b_1 - b_2$ and thus, a or $-a$ would lie properly between $b_1 - b_2$ and 0, which we have just shown to be impossible.

Suppose that $b_1 - b_2$ is archimedean equivalent to $a \in G \setminus \{0\}$, that is, $v_G a = v_G(b_1 - b_2)$. Then by what we have shown, it follows that $v_G a$ is the maximal element of $v_G G$. Consequently, the convex subgroup $\{x \in G \mid v_G x \geq v_G a\}$ is isomorphic to $\{x \in G \mid v_G x \geq v_G a\}/\{x \in G \mid v_G x > v_G a\}$ and hence archimedean. It is an ordered subgroup of the archimedean group $\{x \in H \mid v_G x \geq v_G a\}/\{x \in H \mid v_G x > v_G a\}$ which contains the non-zero image of $b_1 - b_2$. Thus, there is no element of $\{x \in G \mid v_G x \geq v_G a\}$ properly between this image and zero. Since both groups are archimedean, it follows that $\{x \in G \mid v_G x \geq v_G a\}$ must be discretely ordered, and the same is consequently true for G.

Now suppose that $b_1 - b_2$ is not archimedean equivalent to any element of $G \setminus \{0\}$. Then by our initial argument, $v_G(b_1 - b_2) > v_G G$. If $b_2 \neq 0$, this yields $v_G(b_1 - b_2) > v_G b_2$ since $b_2 \in G$, showing that $v_G b_1 = v_G b_2$ and $\mathrm{sign}(b_1) = \mathrm{sign}(b_2)$ by virtue of (V4) and (NV4).

If $(G, <)$ is dense then it is not discretely ordered and the first case is impossible. If $v_G G$ is cofinal in $v_G H$ then the second case is impossible unless $v_G(b_1 - b_2) = \infty$, that is, $b_1 = b_2$. □

We will now apply this lemma to the quasicut (77).

Lemma A.20 *Let $(G, \chi) \subset (H, \chi)$ be an extension of centripetal or centrifugal precontraction groups and assume that (G, χ) is a contraction group. Let $0 \neq b \in H$.*
a) Suppose that $a \in G \setminus \{0\}$ realizes (77). Then $v_G \chi b = v_G a \in v_G G$ and thus, $\chi^2 b = \pm \chi a \in G$.
b) Suppose that $v_G G$ is cofinal in $v_G H$ and $a \in G \setminus \{0\}$ realizes (77). Then $\chi b = a$.
c) If (77) is not realized by any element of G, then $\chi b \notin G$ and the cut induced by χb in G is equal to (77) and thus uniquely determined by the cut (76). The same holds if 0 realizes (77).

Proof Since we assume (G, χ) to be a centripetal or centrifugal contraction group, it is dense by part c) of Lemma A.8, showing that part a) of the foregoing lemma cannot apply to (G, χ). Further, $0 \neq b \in H$ implies $\chi b \neq 0$. Hence, part a) and part b) of our present lemma are direct consequences of the foregoing lemma. The first assertion of c) was already discussed above.

Now assume that 0 realizes (77). Then the foregoing lemma shows that $v_G \chi b = v_G(\chi b - 0) > v_G G$. Since $\chi b \neq 0$, this yields $\chi b \notin G$. Consequently, (77) is the cut induced by χb in $(G, <)$. □

Lemma A.21 *Let $(G, \chi) \subset (H, \chi)$ and $(G, \chi) \subset (H', \chi)$ be two extensions of centripetal or centrifugal contraction groups. Further, let $b \in H$ be such that $\chi^n b \notin G$ for all $n \in \mathbb{N}$. Suppose that some $b' \in H' \setminus \{0\}$ induces in G the same cut as b. Then also $\chi^n b' \notin G$ for all $n \in \mathbb{N}$, and both $(G + \sum_{n \in \mathbb{N}} \mathbb{Z} \chi^n b, \chi)$ and $(G + \sum_{n \in \mathbb{N}} \mathbb{Z} \chi^n b', \chi)$ are precontraction groups. Moreover, there is an isomorphism of these two groups over (G, χ), sending $\chi^n b$ to $\chi^n b'$ for all $n \in \mathbb{N}$.*

Proof We show that the sets $G \cup \{\chi^n b \mid n \in \mathbb{N}\}$ and $G \cup \{\chi^n b' \mid n \in \mathbb{N}\}$, endowed with the restrictions of the orders of H resp. H', are order isomorphic over G by sending $\chi^n b$ to $\chi^n b'$ for every $n \in \mathbb{N}$, and that $v_G \chi^n b \notin v_G G$ for every $n \in \mathbb{N}$. By assumption, b and $b' \neq 0$ induce the same cut in G. Since $\chi^2 b \notin G$, we have $v_G b \notin v_G G$, and part a) of Lemma A.20 shows that $b' \notin G$. We proceed by induction on $n \geq 0$: Suppose that we have already shown that $v_G \chi^n b \notin vG$ and $\chi^n b' \notin G$ and that

$$\mathrm{Cut}_n(b) := (\{\chi^n g \mid G \ni g \leq b\}, \{\chi^n g \mid G \ni g > b\})$$

is the cut induced by both $\chi^n b$ and $\chi^n b'$ in G. Since we know that $\chi^2(\chi^n b) = \chi^{n+2} b \notin G$, it follows that $v_G \chi^{n+1} b \notin v_G G$, and part a) of Lemma A.20 shows that

3. Ordered abelian groups with contractions 129

Cut$_{n+1}(b)$ is not realized by any element of $G \setminus \{0\}$. We have $\chi^n b, \chi^n b' \neq 0$ since $\chi^n b, \chi^n b' \notin G$. Using part c) of Lemma A.20, we may deduce that $\chi^{n+1} b' \notin G$ and that Cut$_{n+1}(b)$ is the cut induced by both $\chi^{n+1} b$ and $\chi^{n+1} b'$ in G. This completes the induction step. It follows that for all $n \in \mathbb{N}$, the order relation between $\chi^n b$ and a given element $g \in G$ is the same as between $\chi^n b'$ and g. Furthermore, the ordering on the sets $\{\chi^n b \mid n \in \mathbb{N}\}$ and $\{\chi^n b' \mid n \in \mathbb{N}\}$ is given by our condition that both (H, χ) and (H', χ) be centripetal or centrifugal contraction groups. Hence, $\chi^n b \mapsto \chi^n b'$ induces an order isomorphism between the ordered sets $G \cup \{\chi^n b \mid n \in \mathbb{N}\}$ and $G \cup \{\chi^n b' \mid n \in \mathbb{N}\}$.

As $\chi^n b' \notin G$ for all $n \in \mathbb{N}$, we have that $v_G \chi^n b' \notin v_G G$ for all $n \in \mathbb{N}$. Hence, the order isomorphism induces an extension of the identity of $v_G G$ to an order isomorphism $\tau : v_G G \cup \{v_G \chi^n b \mid n \in \mathbb{N}\} \to v_G G \cup \{v_G \chi^n b' \mid n \in \mathbb{N}\}$. Moreover, it yields that sign$(\chi^n b)$ = sign$(\chi^n b')$ for all $n \in \mathbb{N}$. By Lemma A.5 it now follows that this isomorphism extends linearly to an order and valuation preserving isomorphism between the groups $G_b := G + \sum_{n \in \mathbb{N}} \mathbb{Z} \chi^n b$ and $G_{b'} := G + \sum_{n \in \mathbb{N}} \mathbb{Z} \chi^n b'$.

On the other hand, (V3) shows that every element in G_b has value either in $v_G G \cup \{\infty\}$ or equal to $v_G \chi^n b$ for some $n \in \mathbb{N}$. Consequently, the group G_b is closed under χ and thus, (G_b, χ) is a precontraction group. Moreover, it follows that the extension of χ from G to G_b is uniquely determined by the cut induced by b in G, together with the condition that $\chi^n b \notin G$ for all $n \in \mathbb{N}$. Since b' induces the same cut as b in G and $\chi^n b' \notin G$ for all $n \in \mathbb{N}$, it now follows that the above isomorphism between G_b and $G_{b'}$ also preserves the precontraction. □

Corollary A.22 *Let $(G, \chi) \subset (H, \chi)$ be an extension of centripetal or centrifugal precontraction groups and assume that (G, χ) is a non-trivial divisible contraction group. Further, let $b \in H$ be such that $\chi^n b \notin G$ for all $n \in \mathbb{N}$. Then (G, χ) is existentially closed in the precontraction group*

$$(G_b, \chi) := (G + \sum_{n \in \mathbb{N}} \mathbb{Z} \chi^n b, \chi) .$$

Proof Let $(G, \chi)^*$ be a $|G|^+$-saturated elementary extension of (G, χ). Then Lemma A.4 shows that there is some non-zero element b' of $(G, \chi)^*$ which induces the same cut as b in G. Hence, the previous lemma gives an embedding of (G_b, χ) in $(G, \chi)^*$ over (G, χ). Now it follows from Lemma A.3 that (G, χ) is existentially closed in (G_b, χ). □

Corollary A.23 *Let $(G, \chi) \subset (H, \chi)$ be an extension of non-trivial divisible centripetal contraction groups. Then there exists a divisible centripetal contraction group $(G', \chi) \subset (H, \chi)$ such that $v_G G'$ is cofinal in $v_G H$ and (G, χ) is existentially closed in (G', χ).*

Proof If $v_G G$ is not cofinal in $v_G H$ then choose $b \in H \setminus \{0\}$ such that $v_G b > v_G G$. Since (H, χ) is centripetal, we have $v_G \chi^n b > v_G G$ for all $n \in \mathbb{N}$. By the foregoing lemma, (G, χ) is existentially closed in the precontraction group (G_b, χ). By Lemma A.18, (G, χ) is also existentially closed in every divisible contraction hull of that group. By its universal property, such a divisible contraction hull can be chosen in (H, χ). If the value set of this new group is still not cofinal in $v_G H$, we may proceed by (possibly transfinite) induction (bounded by the cardinality of $v_G H$) to construct (G', χ) such that $v_G G'$ is cofinal in $v_G H$. □

For centripetal contraction groups, we will now generalize the principle that we have used in the proof of Lemma A.21. The elements b, b' appearing in that lemma seem to be "transcendental" in some sense. For the centripetal case, we will give a corresponding definition.

From now on, all precontraction groups are assumed to be centripetal.

Let $(G, \chi) \subset (H, \chi)$ be an extension of precontraction groups, G divisible, and (H', χ) a substructure of (H, χ) generated over G by one element b. Then as an abelian group, H' is not necessarily generated over G by only the element b. We form a sequence of generators for H' as an abelian group over G in the following way. If $G + \mathbb{Z}b$ is a value set preserving extension of G, then by Lemma A.12, we have $\chi(G + \mathbb{Z}b) \subset G$. In this case, we find $H' = G + \mathbb{Z}b$, and our sequence only consists of the element $b_1 := b$; we set $g_1 := 0$ for later use. In this case, g_1, b_1 and $v_G b_1$ are uniquely determined by b. If on the other hand, the extension is not value set preserving, then there is an element g_1 in the divisible group G such that $v_G(b - g_1) \notin v_G G \cup \{\infty\}$. We set $b_1 := b - g_1$ and find that $G + \mathbb{Z}b = G + \mathbb{Z}b_1$. Note that $v_G b_1$ is uniquely determined by b, by virtue of Lemma A.6. In view of (V3), every element of $G + \mathbb{Z}b_1$ has value either in $v_G G \cup \{\infty\}$ or equal to $v_G b_1$. Hence by (CV1), $\chi(G + \mathbb{Z}b_1) \subset G + \mathbb{Z}\chi b_1 \subset G + \mathbb{Z}b_1 + \mathbb{Z}\chi b_1$. Having constructed b_n and the group $G + \sum_{i=1}^{n} \mathbb{Z}b_i$ such that $v_G b_1, \ldots, v_G b_n$ is a strictly increasing sequence of values not in $v_G G \cup \{\infty\}$ and uniquely determined by b, and such that $\chi(G + \sum_{i=1}^{n} \mathbb{Z}b_i) \subset G + \sum_{i=1}^{n} \mathbb{Z}b_i + \mathbb{Z}\chi b_n$, we proceed as follows.

Since χ is centripetal, we have $v_G \chi b_n > v_G b_n \geq \{v_G b_1, \ldots, v_G b_n\}$. By virtue of (V3), $\chi b_n \in G + \sum_{i=1}^{n} \mathbb{Z}b_i$ will thus imply $\chi b_n \in G$. If this is the case, then $G + \sum_{i=1}^{n} \mathbb{Z}b_i$ is closed under χ, and we let our sequence end with b_n.

If $G + \mathbb{Z}\chi b_n$ is a non-trivial value set preserving extension of G, then we set $b_{n+1} := \chi b_n$ and $g_{n+1} := 0$. Then $v_G b_{n+1} = v_G \chi b_n > v_G b_n$. Note that in this case again, g_{n+1}, b_{n+1} and $v_G b_{n+1}$ are uniquely determined by χb_n. This in turn is, up to the sign, uniquely determined by $v_G b_n$. Hence, $v_G b_{n+1}$ is uniquely determined by $v_G b_n$ and thus by b. The non-zero elements $b_1, \ldots, b_n$ have different values $\notin v_G G = v_G(G + \mathbb{Z}b_{n+1})$ which shows that every element d of $G + \sum_{i=1}^{n+1} \mathbb{Z}b_i$ has value either in $v_G G \cup \{\infty\}$ or equal to $v_G b_i$ for some $i \leq n$. Hence by virtue of (CV3), $\chi(G + \sum_{i=1}^{n+1} \mathbb{Z}b_i) = \chi(G + \sum_{i=1}^{n} \mathbb{Z}b_i) \subset G + \sum_{i=1}^{n+1} \mathbb{Z}b_i$. Consequently, the group $G + \sum_{i=1}^{n+1} \mathbb{Z}b_i$ is closed under χ, and we let our sequence end with b_{n+1}.

If the extension is not value set preserving, then there is an element $g_{n+1} \in G$ such that $v_G(\chi b_n - g_{n+1}) \notin v_G G \cup \{\infty\}$ and we set $b_{n+1} := \chi b_n - g_{n+1}$. We find that

$$v_G b_{n+1} = v_G(\chi b_n - g_{n+1}) \geq v_G \chi b_n > v_G b_n$$

since otherwise, $v_G(\chi b_n - g_{n+1}) = v_G g_{n+1} \in v_G G \cup \{\infty\}$ by (V3). (Note that it is precisely the $\geq$-sign in the foregoing inequality that makes it impossible to construct analogous sequences with strictly decreasing values in centrifugal precontraction groups.) Again by Lemma A.6, $v_G b_{n+1}$ is uniquely determined by χb_n; as before, it follows that $v_G b_{n+1}$ is uniquely determined by b. All elements $b_1, \ldots, b_{n+1}$ have different values $\notin v_G G \cup \{\infty\}$, so every element d of $G + \sum_{i=1}^{n+1} \mathbb{Z}b_i$ has value either in $v_G G \cup \{\infty\}$ or equal to $v_G b_i$ for some $i \leq n+1$. If $v_G d = v_G b_{n+1}$ then $\chi d = \pm \chi b_{n+1}$ by (CV2). Otherwise,

$$\chi d \in \chi\left(G + \sum_{i=1}^{n} \mathbb{Z}b_i\right) \subset G + \sum_{i=1}^{n} \mathbb{Z}b_i + \mathbb{Z}\chi b_n = G + \sum_{i=1}^{n+1} \mathbb{Z}b_i \, .$$

3. Ordered abelian groups with contractions

Hence,
$$\chi\left(G + \sum_{i=1}^{n+1} \mathbb{Z}b_i\right) \subset G + \sum_{i=1}^{n+1} \mathbb{Z}b_i + \mathbb{Z}\chi b_{n+1} \ .$$

Since $v_G b_{n+1} \notin v_G G \cup \{\infty\}$, the procedure will again be repeated.

The sequence of elements b_i constructed in this way may be finite or infinite. If it is finite, the element b will be called **χ-algebraic** over (G, χ), and otherwise, b will be called **χ-transcendental** over (G, χ). These notions are well-defined, although the **shift elements** g_i and the sequence members b_i may not be uniquely determined. Indeed, in our construction, all values $v_G b_i$ were uniquely determined by b, which also yields that all sequences constructed from b must have the same length. Every such sequence will be called a **characteristic sequence** of b over (G, χ). If $b_m \neq 0$ and $v_G b_m > v_G G$ for some m, then the sequence is infinite and $v_G b_i > v_G G$ holds for all $i \geq m$; the sequence beginning with b_m will then be called a **supersequence**. If $v_G G$ is cofinal in $v_G H$, then there are no supersequences in H over (G, χ).

Note that "χ-algebraic" does not mean "algebraic" in the model theoretic sense. We will show in Section 4 (Theorem A.35) that only the elements of the divisible hull $\tilde{G}$ are algebraic over a precontraction group (G, χ).

By construction, the abelian group generated over G by all the elements of a characteristic sequence, is closed under χ and thus a precontraction group. On the other hand, all elements b_i are contained in H' which was the substructure of (H, χ) generated by b. We find that the elements of a characteristic sequence form a set of generators of H' as an abelian group over G. By construction, it is even a minimal set of generators.

For arbitrary b_n in a characteristic sequence, all inclusions in the chain $G \subset G + \mathbb{Z}b_n \subset G + \mathbb{Z}b_n + \mathbb{Z}b_{n-1} \subset \ldots \subset G + \sum_{i=1}^{n} \mathbb{Z}b_i$ are proper. Indeed, from our construction it follows that $v_G b_{j-1} \notin \{v_G b_j, \ldots, v_G b_n\}$ and there is thus no element of value $v_G b_{j-1}$ in the group $G + \sum_{i=j}^{n} \mathbb{Z}b_i$, for every $j \leq n$. Since
$$\chi\left(G + \sum_{i=j}^{n} \mathbb{Z}b_i\right) \subset G + \sum_{i=j+1}^{n} \mathbb{Z}b_i + \mathbb{Z}\chi b_n \ ,$$
we also see that
$$b_j + g \notin \chi\left(G + \sum_{i=j}^{n} \mathbb{Z}b_i\right) \quad \text{for every } j \leq n \text{ and all } g \in G.$$

Hence if $j \geq 2$, then
$$\chi b_{j-1} = b_j + g_{j-1} \notin \chi\left(G + \sum_{i=j}^{n} \mathbb{Z}b_i\right) \ . \tag{79}$$

Note that our assertions also hold for $\mathbb{Q}$ in the place of $\mathbb{Z}$. If $b_1, \ldots, b_n$ is the characteristic sequence of some χ-algebraic element b over (G, χ), then all groups in this chain are precontraction groups since for every $i \leq n$, the sequence $b_i, \ldots, b_n$ is a characteristic sequence of b_i. The construction carried through in the proof of Lemma A.16 may be applied to $(G + \mathbb{Z}b_n, \chi)$ in the place of (G, χ). By virtue of (79), we may take the first adjoined elements to be $b_{n-1}, \ldots, b_1$ successively. In this way, we obtain a contraction hull of $(G + \mathbb{Z}b_n, \chi)$ in (H, χ) which contains all elements of the given characteristic sequence of b. Hence, it also contains the precontraction group generated by b over (G, χ). If H is divisible, then we may replace $\mathbb{Z}$ by $\mathbb{Q}$ to obtain a divisible contraction hull. We have proved:

Lemma A.24 *Let $(G, \chi) \subset (H, \chi)$ be an extension of centripetal precontraction groups and assume that G is divisible and (H, χ) is a contraction group (resp. a divisible contraction group). Let $b \in H$ be χ-algebraic over (G, χ) with characteristic sequence $b_1, \ldots, b_n$. Then there exists a contraction hull (resp. a divisible contraction hull) of $(G + \mathbb{Z}b_n, \chi)$ in (H, χ) which contains the precontraction group generated by b over (G, χ).*

From now on, let us assume (G, χ) to be a divisible centripetal contraction group. Let $b \notin G$. We will now examine the question whether the cuts induced in G by the elements of characteristic sequences of b are already determined by the cut

$$\text{Cut}(b) := (\{g \mid G \ni g < b\}, \{g \mid G \ni g > b\})$$

induced by b in G. To this end, we need the notion of a v_G-cut that we have introduced in Section 2. Bearing in mind the proof of Lemma A.21 where we defined an isomorphism by an assignment $\chi^n b \mapsto \chi^n b'$, we will treat the following problem. Assume that $(G, \chi) \subset (H', \chi)$ is a second extension of centripetal precontraction groups and that $b' \in H'$ also induces the cut $\text{Cut}(b)$ in G. Is it then possible to construct characteristic sequences of b and b' of equal length and having the same shift elements g_i for every i?

Let us fix a characteristic sequence $b_1, \ldots$ of b over (G, χ). For every b_i in this sequence, we have

$$\text{Cut}(b_i) = (\{g \mid G \ni g < b_i\}, \{g \mid G \ni g > b_i\})$$

since by definition, $b_i \notin G$.

Assume that b_m, $m \geq 1$, is an element of our characteristic sequence and that we have already shown $\text{Cut}(b_{m-1})$ to depend only on $\text{Cut}(b)$ (and not on the chosen characteristic sequence or the element $b \notin G$ inducing $\text{Cut}(b)$). For $m = 1$, this assumption is trivially true. If $m > 1$, then assume in addition that we have already constructed the first elements $b'_1, \ldots b'_{m-1}$ of a characteristic sequence of b', satisfying $b'_1 = b' - g_1$ and $b'_i = \chi b_{i-1} - g_i$ as well as $\text{Cut}(b_i) = \text{Cut}(b'_i)$ for $1 \leq i < m$. For the case $m = 1$, we set $C_1 := \text{Cut}(b) = \text{Cut}(b')$. For $m > 1$, we set

$$C_m := (\{\chi g \mid G \ni g < b_{m-1}\}, \{\chi g \mid G \ni g > b_{m-1}\}). \tag{80}$$

Note that a priori, we only know that (80) is a quasicut. But since b_m is an element of our characteristic sequence, $b_m + g_m \notin G$. As $\chi b_{m-1} = b_m + g_m$ realizes (80), this shows that (80) is a cut. Since by induction hypothesis, $\text{Cut}(b_{m-1}) = \text{Cut}(b'_{m-1})$ depends only on $\text{Cut}(b)$, also C_m depends only on $\text{Cut}(b)$, and by virtue of (C$\leq$), both χb_{m-1} and $\chi b'_{m-1}$ realize C_m. In the sequel, let us treat the cases $m = 1$ and $m > 1$ simultaneously by taking the undefined expression "χb_0" to stand for "b".

Assume that C_m is not realized in G. There are two possibilities for C_m. If C_m is not a shifted v_G-cut, then by Corollary A.2, there is no $g \in G$ such that $v_G(\chi b_{m-1} - g) \notin v_G G \cup \{\infty\}$ or $v_G(\chi b'_{m-1} - g) \notin v_G G \cup \{\infty\}$. Then necessarily, $g_m = 0$ and $b_m = \chi b_{m-1}$ as well as $b'_m = \chi b'_{m-1}$ by our construction of characteristic sequences, implying also that $\text{Cut}(b_m) = C_m = \text{Cut}(b'_m)$. Moreover, b_m and b'_m are the last elements of the respective characteristic sequences.

If on the other hand, C_m is a shifted v_G-cut, then by Corollary A.2, there is some $g \in G$ such that $v_G(\chi b_{m-1} - g) \notin v_G G \cup \{\infty\}$. Then by our construction, $v_G b_m = v_G(\chi b_{m-1} - g_m) \notin v_G G \cup \{\infty\}$ which by Corollary A.2 yields that $C_m - g_m$ is the unique v_G-cut among all shifts of C_m. By part b) of Lemma A.1, we find that $v_G(\chi b'_{m-1} - g_m) \notin v_G G \cup \{\infty\}$, so we may set $b'_m := \chi b'_{m-1} - g_m$. But also any other elements $\tilde{b}_m$, $\tilde{b}'_m$ continuing the characteristic sequences would have to satisfy $v_G \tilde{b}_m$, $v_G \tilde{b}'_m \notin v_G G \cup \{\infty\}$ and would thus induce v_G-cuts, which are then equal to $C_m - g_m$ because of its uniqueness. Consequently, $\text{Cut}(b_m) = C_m - g_m = \text{Cut}(b'_m)$ only depends on $\text{Cut}(b)$.

Now assume that C_m is realized by some $a \in G$. Then in view of Lemma A.19, of which only part b) can apply here since G is dense, our construction of characteristic sequences yields $v_G b_m > v_G G$, and b_m is the beginning of a supersequence. Then b is χ-transcendental and for all $i \geq m$, we have $v_G b_i > v_G G$ and all cuts $\text{Cut}_i(b)$ are the same and equal to one of the two possible cuts in G which are realized by 0 (these are $(G^{<0}, \{0\} \cup G^{>0})$ and $(G^{<0} \cup \{0\}, G^{>0})$). But the following pathological case may appear: we cannot exclude the possibility that $\chi b'_{m-1} = a$. In this case, the characteristic sequence of b' would end with b'_{m-1} and b' would be χ-algebraic. Then certainly, the assignment $b \mapsto b'$ would not extend to an isomorphism of precontraction groups over (G, χ). On the other hand, if our characteristic sequence of b does not eventually run into a supersequence, then for none of its elements b_i, the cut $\text{Cut}(b_i)$ is realized in G. In this case, our procedure of constructing b'_i may be accomplished through the full length of the characteristic sequence of b, showing that the characteristic sequences of b' are at least as long as those of b. If symmetrically, we know that also the characteristic sequences of b' do not run into supersequences, then we may conclude that the characteristic sequences of b and b' are of equal length.

If the characteristic sequences of $b \in H$ over (G, χ) are infinite but do not run into supersequences, then b will be called **bounded χ-transcendental** over (G, χ). If $v_G G$ is cofinal in $v_G H$, then there are no supersequences in H over (G, χ), and every χ-transcendental element is already bounded χ-transcendental.

Assume that $b \in H$ is bounded χ-transcendental over (G, χ) and that b' is as above, satisfying $\text{Cut}(b) = \text{Cut}(b')$. Then our construction yields an infinite characteristic sequence $b'_1, \ldots$ of b' with the same shift elements g_i and such that the cuts $\text{Cut}(b_i) = \text{Cut}(b'_i)$ are not realized in G, for all i. In particular, also b' is bounded χ-transcendental over (G, χ). Moreover, we know that all b_i and b'_i have values not contained in $v_G G \cup \{\infty\}$. Let us further note that every isomorphism ι between the ordered groups $(G + \sum_{i=1}^{\infty} \mathbb{Z} b_i, <)$ and $(G + \sum_{i=1}^{\infty} \mathbb{Z} b'_i, <)$ sending b_i to b'_i, will automatically respect the precontraction. Indeed, since every element in the first group has value either in $v_G G \cup \{\infty\}$ or equal to some $v_G b_i$, we only have to check whether $\iota \chi b_i = \chi b'_i$. But this is immediate: $\iota \chi b_i = \iota(b_{i+1} + g_{i+1}) = b'_{i+1} + g_{i+1} = \chi b'_i$.

Now the proofs of the following lemma and corollary are analogous to those of Lemma A.21 and Corollary A.22.

Lemma A.25 *Let $(G, \chi) \subset (H, \chi)$ and $(G, \chi) \subset (H', \chi)$ be two extensions of centripetal contraction groups and assume that G is divisible. Further, let $b \in H$ be bounded χ-transcendental over (G, χ). Suppose that some $b' \in H'$ induces in G the same cut as b. Then also b' is bounded χ-transcendental over (G, χ), and there is an isomorphism of the precontraction groups generated by b resp. b' over (G, χ), sending b to b'.*

Corollary A.26 *Let $(G, \chi) \subset (H, \chi)$ be an extension of centripetal precontraction groups and assume that (G, χ) is a non-trivial divisible contraction group. If (G_b, χ) is the precontraction group generated over G by a bounded χ-transcendental element $b \in H$, then (G, χ) is existentially closed in (G_b, χ).*

Next, we treat the case of extensions generated by one χ-algebraic element.

Lemma A.27 *Let $(G, \chi) \subset (H, \chi)$ be an extension of centripetal precontraction groups and assume that (G, χ) is a non-trivial divisible contraction group. If (H', χ) is a substructure of (H, χ) generated over G by one χ-algebraic element b, then (G, χ) is existentially closed in (H', χ).*

Proof We may assume $b \notin G$ since otherwise, the assertion is trivial. Pick a characteristic sequence $b_1, \ldots, b_n$ of b over (G, χ). Then the group $G + \mathbb{Z}b_n$ is a precontraction group, generated over (G, χ) by b_n. According to Lemma A.24, there is a divisible contraction hull (H_0, χ) of $G + \mathbb{Z}b_n$ in (H, χ) which contains the precontraction group (G_b, χ) generated by b over (G, χ). If we are able to show that (G, χ) is existentially closed in $(G + \mathbb{Z}b_n, \chi)$, then it will follow from Lemma A.18 that (G, χ) is also existentially closed in (H_0, χ) and thus also in (G_b, χ). This shows: w.l.o.g. we may assume from the start that $n = 1$ and $b_1 = b \notin G$. Then the cut induced by b in G is not realized in G, since otherwise, b_1 would be the beginning of a supersequence and b could not be χ-algebraic over (G, χ). We also know that $\chi b \in G$. Since (G, χ) is a contraction group, we may pick some $c \in G$ such that $\chi c = \chi b$. Let us assume that $c < b$; for $c > b$ the proof is analogous. Consider the following set of assertions:

$$\{\text{``}g < x \wedge \chi g = \chi x\text{''} \mid g \in G \wedge c \leq g < b\} \cup \{\text{``}g > x\text{''} \mid g \in G \wedge g > b\}. \quad (81)$$

This set is finitely satisfiable in (G, χ). For this, we only have to show that for every finite subset $\mathcal{F}$ of (81), there is some $x \in G$ for which all assertions of $\mathcal{F}$ are true. We write $\mathcal{F} = \mathcal{F}_1 \cup \mathcal{F}_2$ with $\mathcal{F}_1$ a subset of the first and $\mathcal{F}_2$ a subset of the second set in (81). Let g_1 be the maximal element appearing in the assertions of $\mathcal{F}_1$. Then there is some element $g' \in G$ such that $g_1 < g' < b$ since otherwise, g_1 would realize the cut induced by b in G. Since $c \leq g_1 < g' < b$ and $\chi c = \chi b$, it follows from (CC) that $\chi g = \chi b = \chi g'$ for all g appearing in the assertions of $\mathcal{F}_1$, and for these, we also have $g \leq g_1 < g'$. On the other hand, for every $g > b$ we have $g > g'$. This shows that $x = g'$ satisfies all assertions in $\mathcal{F}$. Thus we have shown that the set (81) is finitely satisfiable in (G, χ). Hence, it is satisfiable in every $|G|^+$-saturated elementary extension $(G, \chi)^*$ of (G, χ). That is, there is some element b' in $(G, \chi)^*$ which induces in G the same cut as b and which satisfies $\chi b' = \chi b$. This yields that the assignment $b \mapsto b'$ defines an order preserving isomorphism from $G + \mathbb{Z}b$ onto $G + \mathbb{Z}b'$ over G. We have to show that this isomorphism is also an isomorphism of precontraction groups. If $G \subset G + \mathbb{Z}b$ is not value set preserving, then by our choice of b (to be the first element of some characteristic sequence of b) we know that $v_G b \notin v_G G \cup \{\infty\}$, and our assertion follows from the uniqueness assertion of Lemma A.9. If $G \subset G + \mathbb{Z}b$ is value set preserving, then the uniqueness assertion of Lemma A.12 shows that every order preserving isomorphism from $G + \mathbb{Z}b$ onto $G + \mathbb{Z}b'$ over G will automatically be an isomorphism of precontraction groups.

We have shown that (G_b, χ) embeds over (G, χ) in every $|G|^+$-saturated elementary extension of (G, χ). Hence, (G, χ) is existentially closed in (G_b, χ), as contended. $\square$

Now we are able to prove our main lemma.

3. Ordered abelian groups with contractions

Lemma A.28 *Let $(G, \chi) \subset (H, \chi)$ be an extension of centripetal precontraction groups and assume that (G, χ) is a non-trivial divisible contraction group. Then (G, χ) is existentially closed in (H, χ).*

Proof In view of Lemma A.15, we may assume w.l.o.g. that (H, χ) is a divisible contraction group. By Corollary A.23, we know that there exists a divisible contraction group (G', χ) in (H, χ) such that $v_G G'$ is cofinal in $v_G H$ and that (G, χ) is existentially closed in (G', χ). Now, it remains to show that (G', χ) is existentially closed in (H, χ). Hence, we may assume from the start that $v_G G$ is cofinal in $v_G H$.

It suffices to show that (G, χ) is existentially closed in every substructure of (H, χ) which is finitely generated over G. Again in view of Lemma A.15, we may thus assume w.l.o.g. that (H, χ) is a divisible contraction hull of a precontraction group which is finitely generated over (G, χ). Picking one of the generators out of a minimal set of, say, m generators, we consider the substructure (G_1, χ) of (H, χ) generated by this element. We know that (G_1, χ) is a precontraction group. If we are able to show that (G, χ) is existentially closed in (G_1, χ), then by Lemma A.16 and Lemma A.18 there is a divisible contraction hull (G'_1, χ) of (G_1, χ) in (H, χ) such that (G, χ) is existentially closed in (G'_1, χ). Now (H, χ) is a divisible contraction hull of a precontraction group which is generated over (G'_1, χ) by $m - 1$ generators. This shows that we may proceed by induction on the number of generators. That is, we only have to show that (G, χ) is existentially closed in every substructure of (H, χ) which is generated over G by one element b. If b is χ-algebraic over (G, χ), then this is the assertion of Lemma A.27. If b is χ-transcendental over (G, χ), then it is bounded χ-transcendental since $v_G G$ is cofinal in $v_G H$, and an application of Lemma A.26 now completes our proof. □

By Robinson's Test, the foregoing lemma implies:

Theorem A.29 *The elementary theory of non-trivial divisible centripetal contraction groups is model complete.*

Now assume that (G, χ) is a common substructure of the two divisible centripetal contraction groups (H, χ) and (H', χ). By Lemma A.7, (G, χ) is a centripetal precontraction group. By Lemma A.16, there is a divisible contraction hull (G', χ) of (G, χ) in (H, χ) which embeds in (H', χ) over (G, χ). Let us identify (G', χ) with its image in (H', χ). By Lemma A.11, the theory of centripetal contraction groups admits $\mathcal{P}_{\text{cp}}$ as its prime structure. Hence if G is the trivial group, then we may replace (G, χ) by $\mathcal{P}_{\text{cp}}$ to obtain that (G', χ) is non-trivial. Now (G', χ) is a non-trivial divisible centripetal contraction group, so from the model completeness stated in the preceding theorem we may infer that (H, χ) and (H', χ) are equivalent over (G', χ) and hence also over (G, χ). We have thus shown that the theory of divisible centripetal contraction groups is substructure complete, that is,

Theorem A.30 *The elementary theory of non-trivial divisible centripetal contraction groups admits elimination of quantifiers in the language $\mathcal{L}_{\text{cg}}$.*

Combining this result with Lemma A.15, we obtain:

Theorem A.31 *The elementary theory of non-trivial divisible centripetal contraction groups is the model completion of the theory of centripetal precontraction groups.*

Since every two centripetal contraction groups contain the trivial group as a common substructure, substructure completeness yields:

Theorem A.32 *The elementary theory of non-trivial divisible centripetal contraction groups is complete.*

Since the axiom system $\{(OAG), (C0), (CS), (C\leq), (C-), (CA'), (CP), (D), \exists x : x \neq 0\}$ is recursive, this theorem implies:

Theorem A.33 *The elementary theory of non-trivial divisible centripetal contraction groups is decidable.*

4 Weak o-minimality

In this section, we shall study the terms built up using contractions and constants from G. (For a detailed study of terms built up without constants, see [KF2]). This leads to a description of the sets which are definable with constants. This description together with Theorem A.30 will show:

Theorem A.34 *The elementary theory of non-trivial divisible centripetal contraction groups is weakly o-minimal.*

A theory (in which an order $<$ is given or definable) is called **weakly o-minimal** if in every model of this theory, every (elementarily) definable subset is a finite union of convex subsets. In contrast to o-minimality, it is not required that these convex subsets be intervals.

Let $\mathcal{M}$ be a model of a given elementary theory $\mathcal{T}$ and A a subset of $\mathcal{M}$. Then an element $a \in \mathcal{M}$ is said to be **algebraic** over A if it satisfies an elementary formula (with constants from A) which is satisfied only by finitely many elements of $\mathcal{M}$. The set of all elements of $\mathcal{M}$ which are algebraic over A is called the **algebraic closure** of A in $\mathcal{M}$. A theory $\mathcal{T}$ is said to have the **algebraic exchange property** if the following holds in every model $\mathcal{M}$ of $\mathcal{T}$: given $A \subseteq \mathcal{M}$ and $a, b \in \mathcal{M}$ such that a lies in the algebraic closure of $A \cup \{b\}$ but not in the algebraic closure of A, then b lies in the algebraic closure of $A \cup \{a\}$. For example, the theory of abelian groups and the theory of fields have the algebraic exchange property. Divisible ordered abelian groups and real closed fields are examples for o-minimal theories having the algebraic exchange property. But the situation may change drastically if we add new symbols to the language from which the formulas are built up. The theory of divisible ordered abelian groups with order-compatible valuation is an example of a weakly o-minimal theory also having the algebraic exchange property. In contrast to this, we have:

Observation. *The elementary theory of divisible centripetal contraction groups does not have the algebraic exchange property.*

This is particularly remarkable since it was the first known naturally appearing weakly o-minimal theory not having the algebraic exchange property. For a different, more "artificial" example, see [MP–M–S]. Note that by a theorem of Pillay and Steinhorn [P–S], every o-minimal theory has the algebraic exchange property. For the above observation, let us give a quick proof which is due to D. Macpherson: Choose an $\aleph_0$-saturated model (G, χ) of the theory. Take $A = \emptyset$ and G_0 to be the algebraic closure of $\emptyset$. Since our language of precontraction groups is finite, $\aleph_0$-saturation yields the existence of some $a \in G \setminus G_0$. Take G_1 to be the algebraic closure of $\{a\}$. Since G_1 is finitely generated and the preimage $\chi^{-1}(a)$ is infinite, $\aleph_0$-saturation yields the existence of some $b \in \chi^{-1}(a) \setminus G_1$. On the other hand,

4. Weak o-minimality

$a = \chi(b)$ shows that a lies in the algebraic closure of $\{b\}$. Since it does not lie in G_0, this witnesses the failure of the algebraic exchange property.

Here, we have not even used the fact that $G_0 = \{0\}$. While this is easy to see, the following will be a non-trivial consequence of our analysis of the terms built up with contractions:

Theorem A.35 *If A is a set of elements in the divisible centripetal contraction group (G, χ), then the algebraic closure of A in G is precisely the divisible hull of the precontraction group generated by A in G.*

Note that in every ordered group, the algebraic closure of a subset A is equal to the definable closure of A. Indeed, in a finite set of algebraic elements, every element can be singled out by means of the order.

From this theorem, one can deduce that the element b of our above example does not even lie in the least convex subgroup of G which contains the algebraic closure of $\{a\}$.

We are now going to show that the theory of divisible centripetal contraction groups is weakly o-minimal. Let us consider a divisible centripetal contraction group (G, χ). By Theorem A.30, its theory admits quantifier elimination in the language $\mathcal{L}_{\text{cg}}$. It follows that every formula in one free variable x and with constants from G is equivalent to a finite boolean combination of formulas of the type $t_1(x) < t_2(x)$ or $t_1(x) = t_2(x)$, where t_1, t_2 are terms in the language $\mathcal{L}_{\text{cg}}(G, x)$. Here, $\mathcal{L}_{\text{cg}}(G, x)$ denotes the language $\mathcal{L}_{\text{cg}}$ augmented by the free variable x and constant names for the elements of G. But since "$-$" is included in $\mathcal{L}_{\text{cg}}$, the above atomic formulas are equivalent to $t_2(x) - t_1(x) > 0$ resp. $t_2(x) - t_1(x) = 0$. So every formula in one free variable x is equivalent to a finite boolean combination of formulas of the type $t(x) > 0$ or $t(x) = 0$. Every term $t(x)$ may be viewed as a map from G into G. It follows that every definable set in a divisible centripetal contraction group (and every quantifier free definable set in a divisible centripetal precontraction group) is a finite boolean combination of sets of the form $t^{-1}(G^{>0})$ and $t^{-1}(\{0\})$. Since $t^{-1}(G^{<0}) = \tau^{-1}(G^{<0})$ where $\tau(x) = t(-x)$, such a finite boolean combination is actually a finite union of finite intersections of sets of the form $t^{-1}(G^{>0})$ and $t^{-1}(\{0\})$. If all sets of the form $t^{-1}(G^{>0})$ and $t^{-1}(\{0\})$ are already finite unions of convex sets, then the same is true for an arbitrary definable set. Hence:

Lemma A.36 *If for every divisible centripetal contraction group (G, χ) and every term $t(x) \in \mathcal{L}_{\text{cg}}(G, x)$, the sets $t^{-1}(G^{>0})$ and $t^{-1}(\{0\})$ are finite unions of convex subsets of G, then the theory of divisible centripetal contraction groups is weakly o-minimal.*

On the other hand, in a contraction group (G, χ) the preimage $\chi^{-1}(a)$ of a non-zero element a can never be an interval. Indeed, by (CV1) and (C$\leq$), such a preimage is of the form $\{g \in G^{>0} \mid v_G g \in \Gamma\}$ if $a > 0$, or $\{g \in G^{<0} \mid v_G g \in \Gamma\}$ if $a < 0$, for a convex subset $\Gamma \subseteq v_G G$. These sets are convex but not intervals: they have no minimal or maximal elements, and the same is true for their complements on the left and on the right. Hence, contraction groups cannot be o-minimal.

In this section, let (G, χ) always be a centripetal precontraction group.

Let us observe that terms containing non-zero constants need not be monotone or surjective. Indeed, consider the map defined by the term

$$t(x) = \chi(x + a) - \chi(x - a)$$

for $0 \neq a \in G$. It is immediately seen to be symmetrical: $t(-g) = t(g)$. For $v_G g < v_G a$, we have $t(g) = \chi g - \chi g = 0$ by (CV3). For $v_G g > v_G a$, we compute $t(g) = \chi a - \chi(-a) = 2\chi a$, again by (CV3). For $v_G g = v_G a$, we have $v_G(g+a) \geq v_G a$ and $v_G(g-a) \geq v_G a$, hence $|\chi(g+a)| \leq |\chi a|$ and $|\chi(g-a)| \leq |\chi a|$ by (CV5), showing that $|t(g)| \leq 2|\chi a|$. So t is neither monotone nor surjective. But we will show that maps defined by terms are piecewise monotone.

A partition of G into finitely many convex subsets will be called a **finite convex partition**. We will say that a map $t: G \to G$ is **piecewise monotone** if there is a finite convex partition of G such that on every single partition set, the map t is monotone. Then it will follow that the preimage of a convex set, if intersected with one of the partition sets, is empty or convex. In particular, the sets $t^{-1}(G^{>0})$ and $t^{-1}(\{0\})$ will be empty or finite unions of convex sets (each partition set containing at most one of them). So in view of Lemma A.36, the weak o-minimality of the theory of divisible centripetal contraction groups will be proved if we are able to show that every divisible centripetal contraction group (G, χ) has the following property:

(PM) *for every term $t(x)$ in the language $\mathcal{L}_{\mathrm{cg}}(G, x)$, the map $x \mapsto t(x)$ is piecewise monotone on G.*

We will determine a class of "polynomials" which represent other terms as maps from suitable convex subsets of G into G. To this end, we introduce generalized χ-polynomials. Beforehand, we define the following convex subgroups of G for $0 \neq a \in G$:

$$\mathcal{O}^a := \{g \in G \mid v_G g \geq v_G a\} \quad \text{and} \quad \mathcal{M}^a := \{g \in G \mid v_G g > v_G a\}.$$

Note that for every $c \in G$, if M is a convex subset of G, then this also holds for

$$c + M = \{c + m \mid m \in M\};$$

in particular, $c + \mathcal{O}^a$ and $c + \mathcal{M}^a$ are convex subsets of G. We also see that for every $v_G a \in v_G G$, the set $\mathcal{O}^a \setminus \mathcal{M}^a = \{g \in G \mid v_G g = v_G a\}$ is the union of two convex sets, namely $\{g \in G \mid v_G g = v_G a \wedge \mathrm{sign}(g) = \mathrm{sign}(a)\}$ and $\{g \in G \mid v_G g = v_G a \wedge \mathrm{sign}(g) = -\mathrm{sign}(a)\}$.

Note that by (CV4), every contraction map is constant on convex sets of the form $a + \mathcal{O}^a$.

Lemma A.37 *Let $a, b, d \in G$.*
a) If $a + \mathcal{M}^d \cap b + \mathcal{M}^d \neq \emptyset$, then $a + \mathcal{M}^d = b + \mathcal{M}^d$.
b) If $v_G(a - b) = \min\{v_G a, v_G b\}$, then $a + \mathcal{M}^a \cap b + \mathcal{M}^b = \emptyset$. If $v_G(a - b) > \min\{v_G a, v_G b\}$, then $a + \mathcal{M}^a = b + \mathcal{M}^b$.

Proof a): Suppose $c \in a + \mathcal{M}^d \cap b + \mathcal{M}^d$. Then $v_G(a - c) > v_G d$ and $v_G(b - c) > v_G d$, hence $v_G(a - b) > v_G d$. Let $b' \in b + \mathcal{M}^d$. Then $v_G(a - b') \geq \min\{v_G(a - b), v_G(b - b')\} > v_G d$, hence $b' \in a + \mathcal{M}_G^{v_G d}$. Symmetrically, one proves $a' \in b + \mathcal{M}^d$ for every $a' \in a + \mathcal{M}^d$. This proves the equality of the two sets.

b): Note that by (V1), $v_G(a - b) \geq \min\{v_G a, v_G b\}$. If $v_G(a - b) = \min\{v_G a, v_G b\}$, say $= v_G a$, then for every $b' \in b + \mathcal{M}^b$ we have $v_G(b - b') > v_G b \geq v_G a$ which yields that $v_G(a - b') = \min\{v_G(a - b), v_G(b - b')\} = v_G a$ by (V3), showing that $b' \notin a + \mathcal{M}^a$. If $v_G(a - b) > \min\{v_G a, v_G b\}$, then by (V4), $v_G a = v_G b$, and we find $b \in a + \mathcal{M}^a$. By part a), this yields equality of the two sets. □

First, we will recursively define **monic generalized χ-monomials** together with their **characteristic domains**:

4. Weak o-minimality

- x and every $c \in G$ are monic generalized χ-monomials, and G is their characteristic domain;
- for every $a \in G$, the term $\chi[a](x) := \chi(x - a)$ is a monic generalized χ-monomial with characteristic domain

$$G[a] := \{g \in G \mid v_G(g - a) > v_G a\} = a + \mathcal{M}^a$$

if $a \neq 0$, and $G[0] := G$ otherwise;

- for $n > 1$ and all $a_1, \ldots, a_n \in G$,

$$\chi[a_1, \ldots, a_n](x) := \chi(\chi[a_1, \ldots, a_{n-1}](x) - a_n)$$

is a monic generalized χ-monomial, provided that its characteristic domain

$$
\begin{aligned}
G[a_1, \ldots, a_n] &:= G[a_1, \ldots, a_{n-1}] \cap \chi[a_1, \ldots, a_{n-1}]^{-1}(G[a_n]) \\
&= \begin{cases} G[a_1, \ldots, a_{n-1}] & \text{if } a_n = 0 \\ \{g \in G[a_1, \ldots, a_{n-1}] \mid v_G(\chi[a_1, \ldots, a_{n-1}](g) - a_n) > v_G a_n\} & \text{if } a_n \neq 0 \end{cases} \\
&= \begin{cases} G[a_1, \ldots, a_{n-1}] & \text{if } a_n = 0 \\ G[a_1, \ldots, a_{n-1}] \cap \chi[a_1, \ldots, a_{n-1}]^{-1}(a_n + \mathcal{M}^{a_n}) & \text{if } a_n \neq 0. \end{cases}
\end{aligned}
\tag{82}
$$

is non-empty.

A term is called a **generalized χ-monomial** if it is of the form $zt(x)$ with $t(x)$ a monic generalized χ-monomial and $0 \neq z \in \mathbb{Z}$, and we define its characteristic domain to be the same as that of $t(x)$.

Given elements $a_1, \ldots, a_n \in G$, we consider

$$f(x) = \sum_{i=1}^{n} z_i \chi[a_1, \ldots, a_i](x) + z_0 x + c \tag{83}$$

with $z_0, \ldots, z_n \in \mathbb{Z}$, $z_n \neq 0$ and $c \in G$. Such a term will be called **generalized χ-polynomial**, provided that $G[a_1, \ldots, a_n]$, called its **characteristic domain**, is non-empty. The sequence $a_1, \ldots, a_n$ will be called the **characteristic sequence** of $f(x)$. The meaning of the characteristic domain is that it is the set of elements where the χ-polynomial, viewed as a map from G into G, cannot be represented in a simpler form. For instance, if $a, g \in G$ such that $v_G(g - a) = v_G a$, then by (CV2), $\chi[a](g) = \chi(g - a) = \pm\chi(a)$; up to the sign (which depends only on the sign of $g - a$), this is a constant not depending on g. If $v_G(g - a) < v_G a$, then $v_G(g - a) = v_G g$, and we have that $\chi[a](g) = \chi(g - a) = \chi(g) = \chi[0](g)$ by virtue of (CV3).

We see at once from the definition that we run into difficulties if we try to add generalized χ-polynomials whose characteristic sequences are different. We will discuss this point later. Beforehand, we have to determine the basic properties of generalized χ-monomials and characteristic domains. We also study preimages of generalized χ-monomials and their intersections with characteristic domains. For the proof of Theorem A.35, we will need the information that such intersections are empty or infinite, apart from a subset of the divisible hull of the precontraction group generated by A. For this purpose, we need the notion of an **open convex set**, which is just a convex set that does not contain a least and does not contain a greatest element. Their intersections are always infinite, if non-empty. There are two basic examples: the open intervals (a, b) in G and the non-trivial convex subgroups of G (or their cosets). We will also view $\emptyset$ as an open convex subset. Note that by our general assumption that (G,χ) be a non-trivial centripetal precontraction group, the value set $v_G G$ has no maximal element, according to Lemma A.8 (note that $\infty \notin v_G G$ by definition). Hence if $d \neq 0$, then the convex subgroup $\mathcal{M}^d$ is non-trivial and thus, every coset $c + \mathcal{M}^d$ is open convex.

A partition of G into finitely many open convex subsets will be called a **finite open convex partition**.

The following lemma states an important property of maps which preserve $\leq$. Its proof is straightforward.

Lemma A.38 *Assume that $(S,<)$ is a totally ordered set and $\chi : S \to S$ is monotone (i.e., χ preserves $\leq$ on all of S or reverses $\leq$ on all of S). Then the preimage of every convex set is convex. If in addition χ is surjective, then the image of every convex set is convex, and the preimage of every open convex set is open convex.*

Here is our first monotonicity result.

Lemma A.39 *Every generalized χ-monomial $z\chi[a_1, \ldots, a_n]$ is a monotone map from G into G; if $z > 0$, it preserves $\leq$, and otherwise, it reverses $\leq$. Under this map, the preimage $(z\chi[a_1, \ldots, a_n])^{-1}(S)$ of a convex set S is convex. If (G, χ) is a centripetal contraction group and $t(x)$ is a monic generalized χ-monomial which is not a constant, then $t(x)$ is surjective and the preimage of an open convex set is again an open convex set. The same holds for every generalized χ-monomial if (G, χ) is a divisible centripetal contraction group.*

Proof For constants, there is nothing to show. Otherwise, our assertion follows by induction: x (the identity map) preserves $\leq$ and is surjective. If $t(x)$ preserves $\leq$, then the same is true for $t(x) - a$ and $\chi(t(x) - a)$, for arbitrary $a \in G$. If $t(x)$ is monotone, then the same is true for $zt(x)$, for every $z \in \mathbb{Z}$. If $t(x)$ is surjective and if (G, χ) is a centripetal contraction group, i.e. χ is surjective, then the same is true for $t(x) - a$ and $\chi(t(x) - a)$, for arbitrary $a \in G$. If in addition G is divisible, then also $zt(x)$ is surjective. The other assertions now follow from Lemma A.38. $\square$

Lemma A.40 *Let $c, a_1, \ldots, a_n \in G$. Assume that $a_1 \neq c$ and set $d = c - a_1$. If $c \in G[a_1, \ldots, a_n]$, then all monomials $\chi[a_1, \ldots, a_i](x)$, $1 \leq i \leq n$, are constant on $c + \mathcal{M}^d$. If $c + \mathcal{M}^d \cap G[a_1, \ldots, a_n] \neq \emptyset$, then $c + \mathcal{M}^d \subseteq G[a_1, \ldots, a_n]$.*

Proof First, suppose that $c \in G[a_1, \ldots, a_n]$ and $c' \in c + \mathcal{M}^d$. Then we have that $v_G(c' - c) > v_G d = v_G(c - a_1)$. Writing $c' - a_1 = (c - a_1) + (c' - c)$, we thus find $\chi(c' - a_1) = \chi(c - a_1)$ by (CV3). Hence, also $\chi[a_1, \ldots, a_i](c') = \chi[a_1, \ldots, a_i](c)$ for $1 \leq i \leq n$. This also yields $c' \in G[a_1, \ldots, a_n]$, showing that $c + \mathcal{M}^d \subseteq G[a_1, \ldots, a_n]$.

Now suppose that $c + \mathcal{M}^d \cap G[a_1, \ldots, a_n] \neq \emptyset$ and that $\tilde{c}$ is an element of this intersection. Then by part a) of Lemma A.37, $c + \mathcal{M}^d = \tilde{c} + \mathcal{M}^d$, and by what we have already proved, we obtain that $c + \mathcal{M}^d \subseteq G[a_1, \ldots, a_n]$. $\square$

From this lemma, we see that the coefficient a_1 plays a special role. For example, we have that $\chi[a_1]^{-1}(\{0\}) = \{a_1\}$. This set is convex but not open. The next lemma shows that only a_1 can prevent a preimage from being open convex:

Lemma A.41 *Assume $n \geq 1$. For every convex subset $S \subseteq G$, the set*

$$\chi[a_1, \ldots, a_n]^{-1}(S) \setminus \{a_1\}$$

is the union of at most two open convex subsets of G. Its intersection with any open convex set not containing a_1 is always an open convex set.

4. Weak o-minimality

If a_1 is not the least or the greatest element of $\chi[a_1,\ldots,a_n]^{-1}(S)$, then this set is open convex.

Also $G[a_1,\ldots,a_n]$ is an open convex set, and if there is a non-zero element among $a_2,\ldots,a_n$, then $a_1 \notin G[a_1,\ldots,a_n]$.

Proof If $S \subseteq G$ is a convex subset of G, then in view of Lemma A.39, the same holds for its preimage $\chi[a_1,\ldots,a_n]^{-1}(S)$. If $c \neq a_1$ is any element of this preimage, then by Lemma A.40, also the open convex coset $c + \mathcal{M}^{(c-a_1)}$ is contained in it. Consequently, $\chi[a_1,\ldots,a_n]^{-1}(S) \setminus \{a_1\}$ is the union of at most two open convex sets. Since $\chi[a_1,\ldots,a_n]^{-1}(S)$ is convex, this shows that its intersection with any open convex set not containing a_1 is again an open convex set. Our arguments also show that only a_1 can be the least or greatest element of $\chi[a_1,\ldots,a_n]^{-1}(S)$; hence if it is not, then this set is open convex.

Now we consider the sets $G[a_1,\ldots,a_n]$. Trivially, G itself is open convex. Let us consider $G[a]$ for $a \in G$. If $a = 0$, then $G[a] = G$ and there is nothing to show. If $a \neq 0$, then
$$G[a] = \{g \in G \mid v_G(g-a) > v_G a\} = a + \mathcal{M}^a$$
which is open convex. Now let $n > 1$ and $a_1,\ldots,a_n \in G$, and assume that we have already shown that $G[a_1,\ldots,a_{n-1}]$ is open convex. If $a_n = 0$, then $G[a_1,\ldots,a_{n-1},0] = G[a_1,\ldots,a_{n-1}]$ by (82) and there is nothing to show. If $a_n \neq 0$, then by (82),
$$G[a_1,\ldots,a_n] = G[a_1,\ldots,a_{n-1}] \cap \chi[a_1,\ldots,a_{n-1}]^{-1}(a_n + \mathcal{M}^{a_n}).$$
Since $a_n + \mathcal{M}^{a_n}$ is convex, both sets on the right-hand side are convex, and consequently, $G[a_1,\ldots,a_n]$ is convex. Suppose that $i > 1$ is the minimal index such that $a_i \neq 0$ (it exists since we have assumed $a_n \neq 0$). Then
$$v_G(\chi[a_1,\ldots,a_{i-1}](a_1) - a_i) = v_G(\chi^{i-1}(0) - a_i) = v_G a_i,$$
showing that a_1 does not lie in $G[a_1,\ldots,a_i]$ and thus also not in $G[a_1,\ldots,a_n]$.

Since $G[a_1,\ldots,a_n]$ is convex, we see that $G[a_1,\ldots,a_{n-1}]$ can only intersect with one of the possibly two open convex subsets which
$$\chi[a_1,\ldots,a_{n-1}]^{-1}(a_n + \mathcal{M}^{a_n}) \setminus \{a_1\}$$
consists of. Since the intersection of open convex subsets of G is again an open convex subset of G, we have now proved that $G[a_1,\ldots,a_n]$ is open convex. $\square$

Lemma A.42 Assume $n \geq m \geq 1$. For every open convex subset $S \subseteq G$,
$$G[a_1,\ldots,a_n] \cap \chi[a_1,\ldots,a_m]^{-1}(S)$$
is an open convex set.

Proof This follows from the foregoing lemma if $a_1 \notin G[a_1,\ldots,a_n]$ or $a_1 \notin \chi[a_1,\ldots,a_m]^{-1}(S)$. So let us assume that $a_1 \in G[a_1,\ldots,a_n] \cap \chi[a_1,\ldots,a_m]^{-1}(S)$. By the foregoing lemma, we find that $n = 1$ or $a_2 = \ldots = a_n = 0$. Thus, $0 = \chi[a_1,\ldots,a_m](a_1) \in S$. Since S is open convex, there is some $b \in G$ such that $b, -b \in S$. If $a_1 \neq 0$, then we may choose b such that $v_G b > v_G a_1$. Then in all cases, $b \in G[a_1] = G[a_1,\ldots,a_n]$. Since χ is centripetal, we find that
$$v_G \chi[a_1,\ldots,a_m](a_1 \pm b) = v_G \chi^m(a_1 \pm b - a_1) = v_G \chi^m(\pm b) > v_G b.$$
Consequently, $|\chi[a_1,\ldots,a_m](a_1 \pm b)| < |b|$ and therefore,
$$\chi[a_1,\ldots,a_m](a_1 \pm b) \in S.$$

This shows that $a_1 + b, a_1 - b \in G[a_1,\ldots,a_n] \cap \chi[a_1,\ldots,a_m]^{-1}(S)$. By the foregoing lemma, it follows that $\chi[a_1,\ldots,a_m]^{-1}(S)$ and hence also $G[a_1,\ldots,a_n] \cap \chi[a_1,\ldots,a_m]^{-1}(S)$ are open convex. $\square$

Lemma A.43 *For all* $g \in G[a_1,\ldots,a_n]$,

$$v_G\chi[a_1,\ldots,a_n](g) > v_G\chi[a_1,\ldots,a_{n-1}](g) > \ldots > v_G\chi[a_1](g)$$
$$> v_G(g - a_1) \geq v_G g.$$

(Read "$\infty > \infty$" as "$\infty \geq \infty$", so that we do not have to exclude the case where some of the monomials are 0.)

Proof Let $g \in G[a_1,\ldots,a_n]$. By definition of this set, $v_G(g - a_1) \geq v_G g$ and $v_G(\chi[a_1,\ldots,a_i](g) - a_{i+1}) \geq v_G\chi[a_1,\ldots,a_i](g)$ for $1 \leq i < n$. Since χ is centripetal, it follows by Lemma A.8 that

$$v_G\chi[a_1,\ldots,a_{i+1}](g) = v_G\chi(\chi[a_1,\ldots,a_i](g) - a_{i+1})$$
$$> v_G(\chi[a_1,\ldots,a_i](g) - a_{i+1}) \geq v_G\chi[a_1,\ldots,a_i](g).$$
$\square$

In contrast to the case without constants, a term $t(x)$ in the language $\mathcal{L}_{\mathrm{cg}}(G,x)$ may not equal a generalized χ-polynomial as a map on all of G. We have to use suitable partitions of G. Given a finite convex partition P of G, we will say that $t(x)$ is P-**representable** if on every set M in P, $t(x)$ equals some generalized χ-polynomial f_M as a map from M to G, and M is contained in the characteristic domain of f_M. Our goal is to show that for every term $t(x)$ in the language $\mathcal{L}_{\mathrm{cg}}(G,x)$ there exists a finite open convex partition P of G such that $t(x)$ is P-representable. This will be done by induction on the complexity of the terms built up in the language $\mathcal{L}_{\mathrm{cg}}(G,x)$. We have to consider addition and the application of "$-$" and "χ".

Let us first consider addition. Assume that t, t' are the terms which we have to add, and that they equal the generalized χ-polynomial f_M, $f'_{M'}$ on the partition sets M, M' respectively. If $M \cap M'$ is non-empty, then the task is to find a finite open convex partition of $M \cap M'$ such that on every of the partition sets, $f_M + f'_{M'}$ equals a generalized χ-polynomial. But $M \cap M'$ is a subset of the intersection of the characteristic domains of f_M and $f'_{M'}$. So first we give such a partition for this intersection and then intersect the partition sets with $M \cap M'$ to obtain a finite open convex partition of $M \cap M'$ (note that the latter set is itself convex). Thus, we can now forget about the sets M, M' and just find the suitable partition of the intersection of the characteristic domains (if this is non-empty). So we have to compute such intersections. Let $a, b \in G$ with $a < b$.

Assume first that $a \neq 0 \neq b$. Then by part b) of Lemma A.37,

$$a + \mathcal{M}^a \cap b + \mathcal{M}^b = \begin{cases} \emptyset & \text{if } v_G(a - b) = \min\{v_G a, v_G b\} \\ a + \mathcal{M}^a = b + \mathcal{M}^b & \text{if } v_G(a - b) > \min\{v_G a, v_G b\}. \end{cases}$$

Hence, $G[a] = G[b]$ if $G[a] \cap G[b] \neq \emptyset$. (CV3), (CV1) and (C–) show that the maps $\chi(x-a)$ and $\chi(x-b)$ behave in the following way:

$\chi(x - b) = \chi(a - b)$ is constant for $x \in a + \mathcal{M}^{(a-b)}$

$\chi(x - a) = \chi(b - a)$ is constant for $x \in b + \mathcal{M}^{(a-b)}$

$\chi(x - a) = \chi(x - b)$ for $x \in a + \mathcal{M}^a$ with $x < a + \mathcal{M}^{(a-b)}$ or $x > b + \mathcal{M}^{(a-b)}$,

$\chi(x - a) = -\chi(x - b)$ for $x \in a + \mathcal{M}^a$ with $a + \mathcal{M}^{(a-b)} < x < b + \mathcal{M}^{(a-b)}$.

4. Weak o-minimality

Note that $a + \mathcal{M}^{(a-b)} < b + \mathcal{M}^{(a-b)}$.

If $a = 0$, $b \neq 0$, then $G[a] \cap G[b] = G \cap G[b] = G[b]$, and $\chi[a](x) = \chi[0](x) = \chi(x)$ is constant on $G[b] = b + \mathcal{M}^b$. If $b = 0$, $a \neq 0$, then $G[a] \cap G[b] = G[a] \cap G = G[a]$, and $\chi[b](x) = \chi[0](x) = \chi(x)$ is constant on $G[a] = a + \mathcal{M}^a$.

More generally, let also $a_1, \ldots, a_m \in G$. Again, assume first that $a \neq 0 \neq b$. Then in view of (82) and part b) of Lemma A.37, we obtain: If $v_G(a - b) = \min\{v_G a, v_G b\}$, then we have that $a + \mathcal{M}_G^{va} \cap b + \mathcal{M}^b = \emptyset$, hence also

$$\chi[a_1, \ldots, a_m]^{-1}(a + \mathcal{M}^a) \cap \chi[a_1, \ldots, a_m]^{-1}(b + \mathcal{M}^b) = \emptyset$$

and thus, $G[a_1, \ldots, a_m, a] \cap G[a_1, \ldots, a_m, b] = \emptyset$. We can neglect this case since we only have to work with *non-empty* intersections of characteristic domains. Now if $v_G(a - b) > \min\{v_G a, v_G b\}$, then we have $a + \mathcal{M}^a = b + \mathcal{M}^b$ and

$$G[a_1, \ldots, a_m, a] = G[a_1, \ldots, a_m, b] \, .$$

We define a finite open convex partition of $G[a_1, \ldots, a_m, a]$ as follows. We set

$$M_2 := G[a_1, \ldots, a_m] \cap \chi[a_1, \ldots, a_m]^{-1}(a + \mathcal{M}^{(a-b)}) \, ,$$
$$M_4 := G[a_1, \ldots, a_m] \cap \chi[a_1, \ldots, a_m]^{-1}(b + \mathcal{M}^{(a-b)}) \, .$$

Since $a + \mathcal{M}^{(a-b)} \subseteq G[a]$ and $b + \mathcal{M}^{(a-b)} \subseteq G[b]$, we have that

$$M_2 \subseteq G[a_1, \ldots, a_m, a] \quad \text{and} \quad M_4 \subseteq G[a_1, \ldots, a_m, b] = G[a_1, \ldots, a_m, a] \, .$$

Since $a + \mathcal{M}^{(a-b)}$ and $b + \mathcal{M}^{(a-b)}$ are open convex subsets of G satisfying

$$a + \mathcal{M}^{(a-b)} < b + \mathcal{M}^{(a-b)} \, ,$$

Lemma A.42 shows that M_2 and M_4 are open convex subsets of G satisfying $M_2 < M_4$. Now we set

$$M_1 := \{g \in G[a_1, \ldots, a_m, a] \mid g < M_2\} \, ,$$
$$M_3 := \{g \in G[a_1, \ldots, a_m, a] \mid M_2 < g < M_4\} \, ,$$
$$M_5 := \{g \in G[a_1, \ldots, a_m, a] \mid g > M_4\} \, .$$

Note that $M_1 < M_2 < M_3 < M_4 < M_5$. By Lemma A.39, the map $\chi[a_1, \ldots, a_m]$ preserves $\leq$. This gives

$M_1 = G[a_1, \ldots, a_m] \cap \chi[a_1, \ldots, a_m]^{-1}(\{g \in G \mid g < a + \mathcal{M}^{(a-b)}\})$,
$M_3 = G[a_1, \ldots, a_m] \cap \chi[a_1, \ldots, a_m]^{-1}(\{g \in G \mid a + \mathcal{M}^{(a-b)} < g < b + \mathcal{M}^{(a-b)}\})$,
$M_5 = G[a_1, \ldots, a_m] \cap \chi[a_1, \ldots, a_m]^{-1}(\{g \in G \mid g > b + \mathcal{M}^{(a-b)}\})$.

Since the three sets on the right hand side are also open convex, we can again infer from Lemma A.42 that M_1, M_3, M_5 are open convex subsets. Thus, $P = \{M_1, M_2, M_3, M_4, M_5\}$ is a finite open convex partition of $G[a_1, \ldots, a_m, a]$.

From our initial discussion of the maps $\chi(x - a)$ and $\chi(x - b)$, we now see that the maps $\chi[a_1, \ldots, a_n, a](x)$ and $\chi[a_1, \ldots, a_m, b](x)$ behave in the following way:

$$\begin{aligned}
\chi[a_1, \ldots, a_m, a](x) &= \chi[a_1, \ldots, a_m, b](x) && \text{for } x \in M_1 \\
\chi[a_1, \ldots, a_m, b](x) &= \chi(a - b) \text{ is constant} && \text{for } x \in M_2 \\
\chi[a_1, \ldots, a_m, a](x) &= -\chi[a_1, \ldots, a_m, b](x) && \text{for } x \in M_3 \\
\chi[a_1, \ldots, a_m, a](x) &= \chi(b - a) \text{ is constant} && \text{for } x \in M_4 \\
\chi[a_1, \ldots, a_m, a](x) &= \chi[a_1, \ldots, a_m, b](x) && \text{for } x \in M_5 \, .
\end{aligned}$$

If $a = 0$, $b \neq 0$, then $G[a_1, \ldots, a_m, a] \cap G[a_1, \ldots, a_m, b] = G[a_1, \ldots, a_m] \cap G[a_1, \ldots, a_m, b] = G[a_1, \ldots, a_m, b]$, and the fact that χ is constant on $b + \mathcal{M}^b$ yields that $\chi[a_1,\ldots,a_m,a](x) = \chi(\chi[a_1,\ldots,a_m](x))$ is constant on $G[a_1,\ldots,a_m,b] = G[a_1,\ldots,a_m] \cap \chi[a_1,\ldots,a_m]^{-1}(b+\mathcal{M}^b)$.

Symmetrically, if $b = 0$, $a \neq 0$, then $G[a_1,\ldots,a_m,a] \cap G[a_1,\ldots,a_m,b] = G[a_1,\ldots,a_m,a] \cap G[a_1,\ldots,a_m] = G[a_1,\ldots,a_m,a]$, and $\chi[a_1,\ldots,a_m,b](x)$ is constant on $G[a_1,\ldots,a_m,a]$.

In both cases, we do not need to partition $G[a_1,\ldots,a_m,a] \cap G[a_1,\ldots,a_m,b]$; on the whole of this open convex intersection, the χ-monomials behave like on M_4 or M_2, respectively.

Now assume that we are given two generalized χ-polynomials. If the characteristic sequence of one is an initial part of that of the other, then we have no difficulty in adding the two polynomials: we only have to add the two constant terms and the coefficients of corresponding monomials. Addition works even in the sense of adding the polynomials as terms. Assume now that we have to add two generalized χ-polynomials

$$f_1(x) = \sum_{i=1}^{n_1} z_i \chi[a_1,\ldots,a_i](x) + z_0 x + c$$

$$f_2(x) = \sum_{i=1}^{n_2} z'_i \chi[b_1,\ldots,b_i](x) + z'_0 x + c'$$

with *non-empty* intersection of their characteristic domains $G[a_1,\ldots,a_{n_1}]$ and $G[b_1,\ldots,b_{n_2}]$. Our goal is to find a finite open convex partition of this intersection such that on each partition set, the sum of $f_1(x)$ and $f_2(x)$ as maps equals another generalized χ-polynomial. Let $m \geq 0$ be the largest integer such that $a_i = b_i$ for all $i \leq m$. We may assume that $m < n_1$ and $m < n_2$ since otherwise, one characteristic sequence is an initial segment of the other, and we are done. Without loss of generality, we may also assume that $a_{m+1} < b_{m+1}$.

Let us assume first that $a_{m+1} = 0$. Then on $G[b_1,\ldots,b_m,b_{m+1}]$ and hence also on $G[a_1,\ldots,a_{n_1}] \cap G[b_1,\ldots,b_{n_2}]$, the monomial $\chi[a_1,\ldots,a_m,a_{m+1}](x)$ and thus also all monomials $z_i \chi[a_1,\ldots,a_m,a_i](x)$ for $m < i \leq n_1$ are constant; thus, their sum is a constant C. Hence on $G[a_1,\ldots,a_{n_1}] \cap G[b_1,\ldots,b_{n_2}]$, the sum $f_1(x)+f_2(x)$ is equal to

$$\sum_{i=m+1}^{n_2} z'_i \chi[b_1,\ldots,b_i](x) + \sum_{i=1}^{m}(z_i + z'_i)\chi[b_1,\ldots,b_i](x) + (z_0 + z'_0)x + c + c' + C.$$

The case of $b_{m+1} = 0$ is symmetrical.

Now we assume that $a \neq 0 \neq b$. Then we have shown above that

$$G[a_1,\ldots,a_m,a_{m+1}] = G[b_1,\ldots,b_m,b_{m+1}]$$

and that this set splits into five open convex sets. Let us see what happens on these sets.

On M_2, $f_2(x)$ has the property that all monomials $\chi[b_1,\ldots,b_i](x)$ are constant for $i > m$. That is, as a map, $f_2(x)$ equals a generalized χ-polynomial $\tilde{f}_2(x)$ with characteristic sequence $b_1,\ldots,b_m$, and we may add $f_1(x)$ and $\tilde{f}_2(x)$ to obtain a map which equals the map $f_1 + f_2$. Note that the characteristic domain of the generalized χ-polynomial $f_1 + \tilde{f}_2$ is that of f_1, containing M_2.

On M_4, now $f_1(x)$ has the property that $f_2(x)$ had on M_2, and the situation is symmetrical. The characteristic domain will be that of f_2, containing M_4.

4. Weak o-minimality

On M_1 and M_5, we can just replace b_{m+1} by a_{m+1} in $f_2(x)$ without changing its behaviour as a map. On M_3 we have to be a bit more careful. The computation

$$\begin{aligned}
\chi[a_1,\ldots,a_m,b_{m+1},b'](x) &= \chi(\chi[a_1,\ldots,a_m,b_{m+1}](x) - b') \\
&= \chi(-\chi[a_1,\ldots,a_m,a_{m+1}](x) - b') \\
&= -\chi[a_1,\ldots,a_m,a_{m+1},-b'](x)
\end{aligned}$$

shows that we have to replace every b_i by $-b_i$ for $i > m+1$, and the coefficients z_i in $f_2(x)$ by $-z_i$ for $i > m$.

On the three sets M_1, M_3 and M_5 we now have turned $f_2(x)$ into a generalized χ-polynomial whose characteristic sequence coincides with that of $f_1(x)$ on the first $m+1$ entries. So we proceed by induction, again partitioning the sets M_1, M_3 and M_5 according to the rules that we have just established. Our induction will stop when the smaller length of the characteristic sequences of f_1 and f_2 is reached (it may as well stop earlier). Note that in every induction step, the characteristic domains obtained always include the intersection of the characteristic domains of f_1 and f_2 and thus all constructed partition sets.

We have proved:

Lemma A.44 *Let $t_1(x)$ and $t_2(x)$ be terms in $\mathcal{L}_{\mathrm{cg}}(G,x)$ and P_1, P_2 finite open convex partitions of G such that $t_i(x)$ is P_i-representable, for $i = 1,2$. Then there is a finite open convex partition P of G which is finer than P_1 and P_2 such that the term $t_1(x) + t_2(x)$ is P-representable.*

Since the application of the function "$-$" turns a generalized χ-polynomial into a generalized χ-polynomial (with the same characteristic domain and characteristic sequence), it now just remains to consider the application of χ. **In what follows, for $n = 0$ we set** $G[a_1,\ldots,a_n] = G$ **and** $\chi[a_1,\ldots,a_n](x) = x$.

Lemma A.45 *a) Let $f(x)$ be a non-zero generalized χ-polynomial of the form (83). Assume that m is the minimal index such that $z_m \neq 0$. Then for every g in the set $\chi[a_1,\ldots,a_m]^{-1}(G[a_{m+1},\ldots,a_n])$ (which contains the characteristic domain $G[a_1,\ldots,a_n]$), the values of all summands $z_i \chi[a_1,\ldots,a_i](g)$ in $f(g)$ with $m+1 \leq i \leq n$ are bigger than $v_G z_m \chi[a_1,\ldots,a_m](g)$ (or equal if $\chi[a_1,\ldots,a_m](g) = 0$). That gives*

$$\begin{aligned}
v_G(f(g) - c) &= v_G(z_m \chi[a_1,\ldots,a_i](g)) \\
\chi(f(g) - c) &= \chi(z_m \chi[a_1,\ldots,a_i](g)) = \mathrm{sign}(z_m) \cdot \chi[a_1,\ldots,a_i,0](g)
\end{aligned}$$

for all $g \in \chi[a_1,\ldots,a_m]^{-1}(G[a_{m+1},\ldots,a_n])$.

b) As a map from its characteristic domain into G, every generalized χ-polynomial is monotone. More precisely, if f and m are as in a), then f is monotonically increasing if $z_m > 0$ and monotonically decreasing if $z_m < 0$. If $m = 0$ then f is strictly monotone. In any case, if S is a convex subset of G then also $G[a_1,\ldots,a_n] \cap f^{-1}(S)$ is convex.

Proof a): If $g \in \chi[a_1,\ldots,a_m]^{-1}(G[a_{m+1},\ldots,a_n])$, then $\chi[a_1,\ldots,a_m](g) \in G[a_{m+1},\ldots,a_n]$ and by virtue of Lemma A.43, all values $v_G \chi[a_1,\ldots,a_i](g)$, for $m+1 \leq i \leq n$, are bigger than $v_G \chi[a_1,\ldots,a_m](g)$ (or equal if $\chi[a_1,\ldots,a_m](g) = 0$). Since multiplication by $z \neq 0$ does not change the value and multiplication by $z = 0$ lifts it to ∞, this proves our first assertion. The formula for $v_G(f(g) - c)$ follows by (V3), and the formula for $\chi(f(g) - c)$ follows by (NV3).

b): We have already shown in Lemma A.39 that every generalized χ-monomial is monotone. Now assume f and m to be as in a). Let $a, b \in G[a_1, \ldots, a_n]$ such that $a \leq b$. We assume that $z_m > 0$ and show that f preserves $\leq$; if $z_m < 0$, one shows in a similar way that f reverses $\leq$. This will prove the monotonicity.

If $\chi[a_1, \ldots, a_m](a) = \chi[a_1, \ldots, a_m](b)$ then $\chi[a_1, \ldots, a_i](a) = \chi[a_1, \ldots, a_i](b)$ for all $i \geq m$, which yields $z_i \chi[a_1, \ldots, a_i](a) = z_i \chi[a_1, \ldots, a_i](b)$ for $0 \leq i \leq n$, and thus, $f(a) = f(b)$.

Now let $\chi[a_1, \ldots, a_m](a) \neq \chi[a_1, \ldots, a_m](b)$. So by $z_m > 0$ and Lemma A.39, we know that $z_m \chi[a_1, \ldots, a_m](a) < z_m \chi[a_1, \ldots, a_m](b)$.

Let us assume first that
$$v_g(z_m \chi[a_1, \ldots, a_m](a) - z_m \chi[a_1, \ldots, a_m](b)) \geq v_g(\chi[a_1, \ldots, a_{m+1}](a)).$$

Since $z_m \neq 0$ and χ is centripetal, this yields that
$$\begin{aligned} v_g(\chi[a_1, \ldots, a_m](a) - \chi[a_1, \ldots, a_m](b)) &= \\ = v_g(z_m \chi[a_1, \ldots, a_m](a) - z_m \chi[a_1, \ldots, a_m](b)) &\geq v_g(\chi[a_1, \ldots, a_{m+1}](a)) \\ = v_G(\chi(\chi[a_1, \ldots, a_m](a) - a_{m+1})) &> v_G(\chi[a_1, \ldots, a_m](a) - a_{m+1}). \end{aligned}$$

Therefore,
$$v_G(\chi[a_1, \ldots, a_m](a) - a_{m+1}) = v_G(\chi[a_1, \ldots, a_m](b) - a_{m+1}),$$
which in turn implies that
$$\begin{aligned} \chi[a_1, \ldots, a_{m+1}](a) &= \chi(\chi[a_1, \ldots, a_m](a) - a_{m+1}) \\ &= \chi(\chi[a_1, \ldots, a_m](b) - a_{m+1}) = \chi[a_1, \ldots, a_{m+1}](b). \end{aligned}$$

It follows that $z_i \chi[a_1, \ldots, a_i](a) = z_i \chi[a_1, \ldots, a_i](b)$ for all $i > m$. Consequently, $f(a) < f(b)$. Similarly, this can be shown if
$$v_g(z_m \chi[a_1, \ldots, a_m](a) - z_m \chi[a_1, \ldots, a_m](b)) \geq v_g(\chi[a_1, \ldots, a_{m+1}](b)).$$

Now let us assume that
$$v_g(z_m \chi[a_1, \ldots, a_m](a) - z_m \chi[a_1, \ldots, a_m](b)) < v_g(\chi[a_1, \ldots, a_{m+1}](a))$$
and that
$$v_g(z_m \chi[a_1, \ldots, a_m](a) - z_m \chi[a_1, \ldots, a_m](b)) < v_g(\chi[a_1, \ldots, a_{m+1}](b)).$$
Then by part a),
$$v_g(z_m \chi[a_1, \ldots, a_m](a) - z_m \chi[a_1, \ldots, a_m](b)) < v_g(z_i \chi[a_1, \ldots, a_i](a))$$
and
$$v_g(z_m \chi[a_1, \ldots, a_m](a) - z_m \chi[a_1, \ldots, a_m](b)) < v_g(z_i \chi[a_1, \ldots, a_i](b))$$
for all $i > m$. Consequently, (NV5) yields that $f(a) < f(b)$. This proves that f preserves $\leq$.

If $m = 0$, that is, if $z_0 \neq 0$, then $a \neq b$ always implies $z_m \chi[a_1, \ldots, a_m](a) \neq z_m \chi[a_1, \ldots, a_m](b)$. Then it follows that the monotonicity is strict.

Finally, the convexity follows from the monotonicity together with Lemma A.38 and from the fact that $G[a_1, \ldots, a_n]$ is convex by virtue of Lemma A.41. $\square$

4. Weak o-minimality

Let f and m be as in the lemma. We will compute $\chi f(g)$ for $g \in G[a_1, \ldots, a_n]$. If the constant term c of f is zero, then the solution is already given by part a) of the lemma. It remains to treat the case where $c \neq 0$. If $f(x) = zx - c$ with c not divisible by z in G, then we cannot expect that χf is a generalized χ-polynomial as defined above. We would have to choose a more general definition such that $\chi(zx - c)$ is also a generalized χ-polynomial. Everything would also work with this definition, but for the sake of simplicity we rather prefer to **require in the following that G is divisible**. So we can put

$$\tilde{c} := -z_m^{-1} c \,.$$

If $m = n$, that is, if $f - c$ is a monomial, then (using (CZ) to remove z_m):

$$\chi f(g) = \text{sign}(z_m) \cdot \chi(\chi[a_1, \ldots, a_m](g) - \tilde{c}) = \text{sign}(z_m) \cdot \chi[a_1, \ldots, a_m, \tilde{c}](g)$$

for all $g \in G[a_1, \ldots, a_m]$, and we are done. If $m < n$, then we proceed as follows. We substitute

$$y(x) := \chi[a_1, \ldots, a_m](x) - a_{m+1}$$

and obtain

$$\left.\begin{aligned} f(x) = F(y) &:= \sum_{i=m+2}^{n} z_i \chi[a_{m+2}, \ldots, a_i](\chi y) + z_{m+1}\chi y + z_m y - z_m d \\ &\text{with} \quad d := -a_{m+1} - z_m^{-1} c = \tilde{c} - a_{m+1} \,. \end{aligned}\right\} \quad (84)$$

We have that $v_G d = v_G(z_m a_{m+1} + c)$.

If $v_G y < v_G d$, then by part a) of Lemma A.45, $z_m y$ is the summand of least value in $F(y)$, and by (CV3), we get $\chi F(y) = \chi(z_m y) = \text{sign}(z_m) \cdot \chi y$. So we obtain

$$\chi f(g) = \text{sign}(z_m) \cdot \chi[a_1, \ldots, a_{m+1}](g) \quad \text{for } g \in G[a_1, \ldots, a_n] \cap y^{-1}(G \setminus \mathcal{O}^d) \,. \quad (85)$$

If $v_G d = \infty$, i.e., $d = 0$, then $\mathcal{O}^d = \{0\}$. In this case, we have that $y(g) = 0$ and $\chi[a_1, \ldots, a_{m+1}](g) = \chi y(g) = 0$ for all $g \in y^{-1}(\mathcal{O}^d)$. Lemma A.43 tells us that then $y \in G[u_1, \ldots, u_n]$ implies that also the higher monomials are zero, so that we find $f(g) = 0$. Hence, the equation of (85) holds for all $g \in G$ in this special case, and we are done.

So let us assume from now on that $d \neq 0$. We are going to split up $G[a_1, \ldots, a_n]$ into five convex subsets. Now $\mathcal{O}^d$ is an open convex subgroup of G, and the set $G \setminus \mathcal{O}^d$ is the union of the two open convex subsets $G^{<0} \setminus \mathcal{O}^d$ and $G^{>0} \setminus \mathcal{O}^d$ of G (one initial and one cofinal). Also the sets $a_{m+1} + (G^{<0} \setminus \mathcal{O}^d)$ and $a_{m+1} + (G^{>0} \setminus \mathcal{O}^d)$ are open convex. By virtue of Lemma A.42, also

$$\begin{aligned} M_1 &:= G[a_1, \ldots, a_n] \cap y^{-1}(G^{<0} \setminus \mathcal{O}^d) \\ M_5 &:= G[a_1, \ldots, a_n] \cap y^{-1}(G^{>0} \setminus \mathcal{O}^d) \end{aligned}$$

are open convex. Moreover,

$$M' := G[a_1, \ldots, a_n] \cap y^{-1}(\mathcal{O}^d)$$

is open convex, and $\{M_1, M', M_5\}$ is a finite open convex partition of $G[a_1, \ldots, a_n]$ such that $M_1 < M' < M_5$.

If $v_G y = v_G(y - d) = v_G d$, then by part a) of Lemma A.45,

$$\chi F(y) = \chi(z_m y + z_m a_{m+1} + c) = \text{sign}(z_m) \cdot \chi(y - d) = \text{sign}(z_m) \cdot (\pm \chi d) \,,$$

and with $\epsilon := \text{sign}(z_m) \cdot \chi|d|$, we find that

$$\chi f(g) = \begin{cases} -\epsilon & \text{on } M_2 := G[a_1, \ldots, a_n] \cap y^{-1}(\{g \in \mathcal{O}^d \mid g < d + \mathcal{M}^d\}) \\ \epsilon & \text{on } M_4 := G[a_1, \ldots, a_n] \cap y^{-1}(\{g \in \mathcal{O}^d \mid g > d + \mathcal{M}^d\}) \,. \end{cases} \quad (86)$$

Let us set
$$M_3 := G[a_1,\ldots,a_n] \cap y^{-1}(d+\mathcal{M}^d)$$
$$= G[a_1,\ldots,a_n] \cap \chi[a_1,\ldots,a_m]^{-1}(\tilde{c}+\mathcal{M}^d).$$

Since
$$\{\{g \in \mathcal{O}^d \mid g < d+\mathcal{M}^d\}, d+\mathcal{M}^d, \{g \in \mathcal{O}^d \mid g > d+\mathcal{M}^d\}\}$$
is a finite open convex partition of $\mathcal{O}^d$, Lemma A.42 shows that $\{M_2, M_3, M_4\}$ is a finite open convex partition of M' satisfying $M_2 < M_3 < M_4$. We find that $\{M_1, M_2, M_3, M_4, M_5\}$ is a finite open convex partition of $G[a_1,\ldots,a_n]$, satisfying $M_1 < M_2 < M_3 < M_4 < M_5$. On the sets M_1 and M_5, $\chi f(g)$ equals the generalized χ-monomial $\mathrm{sign}(z_m) \cdot \chi[a_1,\ldots,a_{m+1}](x)$, according to (85). Note that M_1 and M_5 are contained in $G[a_1,\ldots,a_n]$, which in turn is a subset of the characteristic domain $G[a_1,\ldots,a_{m+1}]$ of this χ-monomial. On the sets M_2 and M_4, $\chi f(g)$ equals the constant $-\epsilon$ resp. ϵ, according to (86). Note that M_2 and M_4 are subsets of the characteristic domain G of the constants $-\epsilon$ and ϵ.

Finally, it remains to analyze the behaviour of χf on M_3, or in other words, for $y \in d + \mathcal{M}^d$. Fortunately, in this case we know the behaviour of the higher monomials very well: by virtue of (NV3), we have $\chi y = \chi d = \chi[a_{m+1}](-z_m^{-1}c) = \chi[a_{m+1}](\tilde{c})$ and thus also
$$\chi[a_{m+2},\ldots,a_i](\chi y) = \chi[a_{m+2},\ldots,a_i](\chi d) = \chi[a_{m+1},\ldots,a_i](\tilde{c})$$
for $m+2 \leq i \leq n$. That is, if we set
$$c' = f^{[m]}(\tilde{c}) \quad \text{with} \quad f^{[m]}(x) := x - z_m^{-1} \cdot \sum_{i=m+1}^{n} z_i \chi[a_{m+1},\ldots,a_i](x), \qquad (87)$$
then for all $g \in M_3$ we have
$$f(g) = z_m \cdot (y + a_{m+1} - c'),$$
$$\chi f(g) = \mathrm{sign}(z_m) \cdot \chi(y + a_{m+1} - c') = \mathrm{sign}(z_m) \cdot \chi[a_1,\ldots,a_m,c'](g).$$

Note that by part b) of Lemma A.45, the generalized χ-polynomial $f^{[m]}(x)$ is strictly monotone on its characteristic domain $G[a_{m+1},\ldots,a_n]$.

Now it remains to show that the partition set M_3 is contained in the characteristic domain $G[a_1,\ldots,a_m,c']$ of the χ-monomial $\mathrm{sign}(z_m) \cdot \chi[a_1,\ldots,a_m,c'](g)$. For later use, let us prove a little bit more. Assume that
$$\tilde{c} + \mathcal{M}^d \cap G[a_{m+1},\ldots,a_n] \neq \emptyset. \qquad (88)$$
Then
$$v_G(\tilde{c} - c') > v_G d = v_G(\tilde{c} - a_{m+1}) > v_G a_{m+1} = v_G \tilde{c} = v_G c',$$
and it follows that $\tilde{c}, c' \in G[a_{m+1},\ldots,a_n]$. Indeed, let us assume (88). Then by Lemma A.40, $\tilde{c} + \mathcal{M}^d \subseteq G[a_{m+1},\ldots,a_n]$. In particular, $\tilde{c} \in G[a_{m+1},\ldots,a_n]$, and from Lemma A.43 we deduce that $v_G \chi[a_{m+1},\ldots,a_i](\tilde{c}) > v_G \chi(\tilde{c} - a_{m+1}) > v_G(\tilde{c} - a_{m+1}) = v_G d > v_G a_{m+1} = v_G \tilde{c}$ for $m+1 \leq i \leq n$. Since $v_G \tilde{c} = v_G c$ and multiplication by z_m does not change the value, it follows from the definition of c' that $v_G(\tilde{c} - c') > v_G d$. Hence, $v_G c' = v_G \tilde{c}$ and $c' + \mathcal{M}^d = \tilde{c} + \mathcal{M}^d$, showing that $c' \in G[a_{m+1},\ldots,a_n]$.

As a special case, we obtain the fact that $\tilde{c} \in G[a_{m+1},\ldots,a_n]$ implies that $c' \in G[a_{m+1},\ldots,a_n]$. That is, $f^{[m]}$ maps $G[a_{m+1},\ldots,a_n]$ into $G[a_{m+1},\ldots,a_n]$.

4. Weak o-minimality

Further, if $M_3 \neq \emptyset$, then (88) holds, and by what we have just proved, we get

$$\begin{aligned} M_3 &= G[a_1,\ldots,a_n] \cap \chi[a_1,\ldots,a_m]^{-1}(\tilde{c} + \mathcal{M}^d) \\ &= G[a_1,\ldots,a_n] \cap \chi[a_1,\ldots,a_m]^{-1}(c' + \mathcal{M}^d) \\ &\subseteq G[a_1,\ldots,a_n] \cap \chi[a_1,\ldots,a_m]^{-1}(c' + \mathcal{M}^{c'}) \subseteq G[a_1,\ldots,a_m,c'] \,. \end{aligned}$$

We have shown that on all partition sets of the finite open convex partition $\{M_1, M_2, M_3, M_4, M_5\}$ of $G[a_1, \ldots, a_n]$, the map χf equals a suitable χ-monomial (which may even be a constant), such that the partition set is contained in the characteristic domain of the χ-monomial. As before in the case of addition, we have to intersect the above partition with the open convex set $M \subseteq G[a_1, \ldots, a_n]$ on which f was given, obtaining a new finite open convex partition. We have proved:

Lemma A.46 *Let $t(x)$ be a term in $\mathcal{L}_{\mathrm{cg}}(G, x)$ and P a finite open convex partition of G such that $t(x)$ is P-representable. Then there is a finite open convex partition P' of G which is finer than P such that the term $\chi t(x)$ is P'-representable.*

In view of Lemma A.36, the following lemma proves Theorem A.34.

Lemma A.47 *For all divisible centripetal precontraction groups (G, χ) and every term $t(x)$ in the language $\mathcal{L}_{\mathrm{cg}}(G, x)$, there is a finite open convex partition P of G such that $t(x)$ is P-representable. Every divisible centripetal precontraction group has the property (PM).*

Proof The proof proceeds by induction on the complexity of the terms built up in the language $\mathcal{L}_{\mathrm{cg}}(G, x)$. With $P = \{G\}$, our assertion is true for all terms $zx + c$ with $z \in \mathbb{Z}$ and $c \in G$. Next, changing t to $-t$ needs no change of the partition. The addition of two terms was treated in Lemma A.44. Finally, the application of χ to a term was treated in Lemma A.46.

The second assertion follows from the first by virtue of part b) of Lemma A.45. □

Let us describe the sets of roots of generalized χ-polynomials f in a divisible centripetal contraction group (G, χ). We start with two special cases that we derive from our above considerations. If $f - c$ is a generalized χ-monomial, say $z_n \chi[a_1, \ldots, a_n](x)$, then the set of roots of f is precisely

$$\chi[a_1, \ldots, a_n]^{-1}(\tilde{c}), \quad \text{where } \tilde{c} = -z_n^{-1} c \,.$$

If $f - c = z_0 x$, then the only root of f is $-z_0^{-1} c$, which lies in the divisible hull of the subgroup generated by c.

Now let f be represented in the form (84) with $m < n$. We consider the special case of $d = 0$, which means that $\tilde{c} = a_{m+1}$. From part a) of Lemma A.45 we conclude that every root

$$g \in \chi[a_1, \ldots, a_m]^{-1}(G[a_{m+1}, \ldots, a_n])$$

of f must satisfy $y(g) = 0$ and must therefore lie in the set

$$y^{-1}(\{0\}) = \chi[a_1, \ldots, a_m]^{-1}(a_{m+1}) \,. \tag{89}$$

But

$$\chi[a_1, \ldots, a_m]^{-1}(G[a_{m+1}, \ldots, a_n]) \cap \chi[a_1, \ldots, a_m]^{-1}(a_{m+1}) \neq \emptyset$$

if and only if $a_{m+1} \in G[a_{m+1}, \ldots, a_n]$ which by Lemma A.41 holds if and only if $n = m+1$ or $a_i = 0$ for $m+1 < i \leq n$. In both cases, all higher terms are 0 if $y(g) = 0$, so $y(g) = 0$ implies that $f(g) = 0$. This proves that f has a root on $\chi[a_1, \ldots, a_m]^{-1}(G[a_{m+1}, \ldots, a_n])$ if and only if

$$\tilde{c} = a_{m+1} \in G[a_{m+1}, \ldots, a_n] \, .$$

In this case, the set of roots of f on $\chi[a_1, \ldots, a_m]^{-1}(G[a_{m+1}, \ldots, a_n])$ is just the set (89). If c' is defined as in (87), then $c' = \tilde{c} = a_{m+1}$ because all higher monomials are zero. So we can write the set (89) as

$$\chi[a_1, \ldots, a_m]^{-1}(c') \, .$$

Formally, one obtains the same for the case where $f - c$ is a generalized χ-monomial since then, the sum over the higher monomials in (87) is empty. With the convention that $G[a_{n+1}, \ldots, a_n] = G$, the above condition on $\tilde{c}$ for f to have a root also holds in this case, because $\tilde{c} \in G$ always holds and f always has a root.

The following lemma shows that our results also hold in general.

Lemma A.48 *Let (G, χ) be a divisible centripetal contraction group, and let f and m be as in Lemma A.45. Then the equation $f(x) = 0$ has a solution in the set*

$$\chi[a_1, \ldots, a_m]^{-1}(G[a_{m+1}, \ldots, a_n]) \tag{90}$$

if and only if $\tilde{c} = -z_m^{-1} c \in G[a_{m+1}, \ldots, a_n]$. If this is the case, then with c' as defined in (87), the set of solutions in the set (90) is precisely the non-empty convex set

$$\chi[a_1, \ldots, a_m]^{-1}(c') \, .$$

Proof We can assume that $m < n$ and $c \neq -z_m a_{m+1}$ because we have already treated the remaining cases. Let f be represented in the form (84). By our assumption, $d \neq 0$.

Suppose that $\tilde{c} \in G[a_{m+1}, \ldots, a_n]$ and let $d = \tilde{c} - a_{m+1}$. We have already shown that this implies $c' \in \tilde{c} + \mathcal{M}^d$ and $c' \in G[a_{m+1}, \ldots, a_n]$. But if this holds, then there is some a in the set (90) such that $\chi[a_1, \ldots, a_m](a) = c'$. Indeed, since we have assumed (G, χ) to be a divisible centripetal contraction group, we know by Lemma A.39 that every generalized χ-monomial is surjective (if it is not a constant). Using the fact that the monomials $\chi[a_{m+1}, \ldots, a_i](x)$ are constant on $\tilde{c} + \mathcal{M}^d$, we obtain:

$$\begin{aligned}
f(a) &= c + z_m \chi[a_1, \ldots, a_m](a) + \sum_{i=m+1}^{n} z_i \chi[a_1, \ldots, a_i](a) \\
&= c + z_m c' + \sum_{i=m+1}^{n} z_i \chi[a_{m+1}, \ldots, a_i](c') \tag{91} \\
&= c + z_m c' + \sum_{i=m+1}^{n} z_i \chi[a_{m+1}, \ldots, a_i](\tilde{c}) = c + z_m c' - z_m c' - c = 0 \, .
\end{aligned}$$

We have thereby shown that every $a \in \chi[a_1, \ldots, a_m]^{-1}(c')$ satisfies $f(a) = 0$. Since $c' \in G[a_{m+1}, \ldots, a_n]$, the set $\chi[a_1, \ldots, a_m]^{-1}(c')$ is a subset of (90). It is non-empty: indeed, since we assume that (G, χ) is a divisible centripetal contraction group, $\chi[a_1, \ldots, a_m]$ is surjective.

For the converse, assume that there is some a in the set (90) such that $f(a) = 0$. Set $c_a := \chi[a_1, \ldots, a_m](a) \in G[a_{m+1}, \ldots, a_n]$. Then in view of Lemma A.43,

$$v_G(z_m c_a + c) = v_G\left(\sum_{i=m+1}^{n} z_i \chi[a_{m+1}, \ldots, a_i](c_a)\right)$$
$$\geq v_G \chi[a_{m+1}](c_a) > v_G(c_a - a_{m+1})$$

and thus, $v_G(c_a - \tilde{c}) = v_G(c_a + z_m^{-1} c) > v_G(c_a - a_{m+1}) = v_G(\tilde{c} - a_{m+1})$. This shows that $c_a \in \tilde{c} + \mathcal{M}^d$. By Lemma A.40 it also follows that $\tilde{c} \in G[a_{m+1}, \ldots, a_n]$. It remains to show that $a \in \chi[a_1, \ldots, a_m]^{-1}(c')$. Using again the fact that the monomials $\chi[a_{m+1}, \ldots, a_i](x)$ are constant on $\tilde{c} + \mathcal{M}^d$, we compute:

$$0 = f(a) = c + z_m \chi[a_1, \ldots, a_m](a) + \sum_{i=m+1}^{n} z_i \chi[a_1, \ldots, a_i](a)$$
$$= c + z_m c_a + \sum_{i=m+1}^{n} z_i \chi[a_{m+1}, \ldots, a_i](c_a)$$
$$= c + z_m c_a + \sum_{i=m+1}^{n} z_i \chi[a_{m+1}, \ldots, a_i](\tilde{c}).$$

It follows that

$$c_a = \tilde{c} - z_m^{-1} \sum_{i=m+1}^{n} z_i \chi[a_{m+1}, \ldots, a_i](\tilde{c}) = c',$$

showing that $a \in \chi[a_1, \ldots, a_m]^{-1}(c')$. We have now shown that $\chi[a_1, \ldots, a_m]^{-1}(c')$ is the set of roots of f in the set (90), provided that there is any root in this set. □

Assume f and m to be as in Lemma A.45. Given $b \in G$, the foregoing lemma applied to $f - b$ helps us to determine the convex set

$$S_f(b) := G[a_1, \ldots, a_n] \cap f^{-1}(b).$$

Indeed, with c'_b determined by (87), applied to $f - b$ in the place of f, we find that

$$S_f(b) = G[a_1, \ldots, a_n] \cap \chi[a_1, \ldots, a_m]^{-1}(c'_b).$$

Assume that $n = 0$. Then $G[a_1, \ldots, a_n] = G$ is open convex, and the unique root of $f - b = z_0 x + c - b$ is $c'_b = -z_0^{-1}(c - b)$. In particular, if $c \in A$, we find that $f^{-1}(\{0\})$ is contained in the divisible hull of the subgroup generated by A. Further, for $b = 0$ we obtain that $f^{-1}(G^{>0}) = z_0^{-1}(G^{>0} - c) = -z_0^{-1} c + \text{sign}(z_0) G^{>0} = c' + \text{sign}(z_0) G^{>0}$, which is open convex.

Now assume that $n > 0$. By Lemma A.41, $G[a_1, \ldots, a_n]$ is open convex, and $\chi[a_1, \ldots, a_m]^{-1}(c'_b) \setminus \{a_1\}$ is the union of at most two open convex sets. Consequently, also $S_f(b) \setminus \{a_1\}$ is the union of at most two open convex sets. In particular, if $a_1 \in A$, then we can say that $S_f(b)$ is a finite union of open convex sets and subsets of A. It follows that

$$G[a_1, \ldots, a_n] \cap f^{-1}(G^{>0}) = \bigcup_{b \in G^{>0}} S_f(b)$$

is a (possibly infinite) union of open convex sets and subsets of A.

In all cases, we may conclude that *if all coefficients of f lie in A, then the sets $G[a_1, \ldots, a_n] \cap f^{-1}(\{0\})$ and $G[a_1, \ldots, a_n] \cap f^{-1}(G^{>0})$ are (possibly infinite)* unions of open convex sets and subsets of the divisible hull of the subgroup generated by A.

Recall that an element a is called definable (or quantifier free definable) if the singleton $\{a\}$ is definable (or quantifier free definable). The definable closure of a set A is the set of all elements which are definable with constants from A. Using the above results, we prove:

Theorem A.49 *Let (G, χ) be a divisible centripetal precontraction group, and $A \subset G$. Then the set of elements which are quantifier free definable in the language $\mathcal{L}_{\mathrm{cg}}$ with constants from A, is equal to the divisible hull of the precontraction group generated by A in G.*

Proof The terms in the simple language of ordered groups with constants from a set A already define all elements of the divisible hull of the subgroup generated by A. The application of χ defines all elements in the precontraction group generated by A. Hence, all elements of the divisible hull of the precontraction group generated by A are definable, using constants from A. So it suffices to show that the definable closure of a set A is contained in the divisible hull of the precontraction group generated by A. Without loss of generality, we may replace A by the divisible hull of the precontraction group generated by A. Then by Lemma A.12, A is itself a precontraction group; so the divisible hull of the precontraction group generated by A is just A.

In our considerations leading to Lemma A.36, we have seen that every set which is quantifier free definable with constants from G, is a finite union of finite intersections of sets of the form $t^{-1}(G^{>0})$ and $t^{-1}(\{0\})$. If the set is already quantifier free definable with constants from A, then also the terms t can be built up using only constants from A. Suppose we can show that for such t, the sets $t^{-1}(G^{>0})$ and $t^{-1}(\{0\})$ are (possibly infinite) unions of open convex sets and subsets of A. Since non-empty intersections of open convex sets are always infinite, it will then follow that every finite set definable with constants from A will already be contained in A, which yields the desired result.

By Lemma A.47 there is a finite open convex partition P of G such that $t(x)$ is P-representable. For $M \in P$, let t be represented by f_M on M (which is a subset of the characteristic domain of f_M). By our proof of Lemma A.47 it is clear that f_M can be constructed from t using only constants in A.

From our above discussion we see that if $G[a_1, \ldots, a_n]$ is the characteristic domain of f_M, then $G[a_1, \ldots, a_n] \cap f_M^{-1}(\{0\})$ and $G[a_1, \ldots, a_n] \cap f_M^{-1}(G^{>0})$ are unions of open convex sets and subsets of A. Thus, also the sets $M \cap f_M^{-1}(\{0\}) = M \cap G[a_1, \ldots, a_n] \cap f_M^{-1}(\{0\})$ and $M \cap f_M^{-1}(G^{>0}) = M \cap G[a_1, \ldots, a_n] \cap f_M^{-1}(G^{>0})$ are unions of open convex sets and subsets of A. We conclude that the sets

$$t^{-1}(\{0\}) = \bigcup_{M \in P} M \cap f_M^{-1}(\{0\}) \quad \text{and} \quad t^{-1}(G^{>0}) = \bigcup_{M \in P} M \cap f_M^{-1}(G^{>0})$$

are unions of open convex sets and subsets of A. This completes our proof. □

Now we can give the

Proof of Theorem A.35. Since the theory of non-trivial divisible centripetal contraction groups admits quantifier elimination (Theorem A.30), every set which is definable in the language $\mathcal{L}_{\mathrm{cg}}$ with constants from A, is also quantifier free definable in the language $\mathcal{L}_{\mathrm{cg}}$ with constants from A. Hence we can infer from the foregoing

4. Weak o-minimality

theorem that the definable closure of a subset A in a divisible centripetal contraction group G is equal to the divisible hull of the precontraction group generated by A in G. Since for ordered structures, definable closure is the same as algebraic closure, this proves Theorem A.35.

Our remark succeeding to Theorem A.35 is proved as follows. Since (G,χ) is centripetal, $\chi b = a$ implies that $v_G b < v_G a$. For the same reason, all elements in the precontraction group H generated by a have value $\geq v_G a$, and the same is true for its divisible hull $\tilde H$ because it has the same value set as H. By definition of v_G, the same is also true for the convex hull of $\tilde H$ in G. Consequently, b is not an element of this convex hull.

For the conclusion of this appendix, we derive two corollaries from Lemma A.48. The first is a criterion for roots which resembles a bit the Hensel-Rychlik criterion for roots of polynomials over henselian valued fields (because the bound does not depend on the approximative root).

Corollary A.50 *Let (G,χ) be a divisible centripetal contraction group, and let f and m be as in Lemma A.45. Assume that $m < n$ and $c \neq -z_m a_{m+1}$. Then the equation $f(x) = 0$ has a solution in the set (90) if and only if*
$$\exists x \in \chi[a_1,\ldots,a_m]^{-1}(G[a_{m+1},\ldots,a_n]): v_G f(x) > v_G(\tilde c - a_{m+1}).$$

Proof If $f(a) = 0$ then trivially, $v_G f(a) > v_G(\tilde c - a_{m+1})$. For the converse, suppose that $v_G f(b) > v_G(\tilde c - a_{m+1})$ for some b in the set (90). We set $c_b := \chi[a_1,\ldots,a_m](b)$. If $v_G f(b) = v_G(c_b - \tilde c)$ then we obtain that
$$v_G(c_b - \tilde c) > v_G(\tilde c - a_{m+1}) = v_G(c_b - a_{m+1}).$$
On the other hand, if $v_G f(b) \neq v_G(c_b - \tilde c)$ then there must be some $\imath > m$ such that $v_G(c_b - \tilde c) \geq v_G z_\imath \chi[a_{m+1},\ldots,a_\imath](c_b)$ which in view of $c_b \in G[a_{m+1},\ldots,a_n]$ and Lemma A.43 yields that $v_G(c_b - \tilde c) > v_G(c_b - a_{m+1})$. We have shown that the latter holds in every case. It follows from Lemma A.40 that $\tilde c \in G[a_{m+1},\ldots,a_n]$, which by Lemma A.48 yields the existence of the required root. □

Finally, let us show that a generalized χ-polynomial preserves convexity on its characteristic domain:

Corollary A.51 *Let (G,χ) be a divisible centripetal contraction group. If f is a generalized χ-polynomial and M is a convex subset of its characteristic domain, then $f(M)$ is a convex subset of G.*

Proof Let f and m be as in Lemma A.45. Suppose that there are $g_1, g_2 \in M$ and $b \in G$ such that $f(g_1) < b < f(g_2)$. We want to show that there is $g_b \in M$ such that $f(g_b) = b$. Since $f - f(g_i)$, $i = 1, 2$, have a zero in M, and since M is a subset of $G[a_1,\ldots,a_n]$ and hence also of the set (90), Lemma A.48 shows that $\tilde c_i := -z_m^{-1}(c - f(g_i)) \in G[a_{m+1},\ldots,a_n]$ and that $g_i \in \chi[a_1,\ldots,a_m]^{-1}(c'_i)$, where $c'_i := f^{[m]}(\tilde c_i)$ as defined in (87). The element $\tilde c_b := -z_m^{-1}(c - b)$ lies properly between $\tilde c_1$ and $\tilde c_2$ and since $G[a_{m+1},\ldots,a_n]$ is convex, it is also contained in $G[a_{m+1},\ldots,a_n]$. Thus, setting $c'_b := f^{[m]}(\tilde c_b)$, we deduce from Lemma A.48 that all elements of the set $\chi[a_1,\ldots,a_m]^{-1}(c'_b)$ are solutions for $f(x) - b = 0$. We also find that c'_b lies properly between c'_1 and c'_2 since $f^{[m]}$ is strictly monotone. Since $\chi[a_1,\ldots,a_m]$ is monotone, it follows that all elements of $\chi[a_1,\ldots,a_m]^{-1}(c'_b)$ lie between g_1 and g_2 and are thus contained in M because M was assumed to be convex. We have shown that the equation $f(x) = b$ admits a solution in M. □

Bibliography

[A] Alling, N. L.: *On exponentially closed fields*, Proc. Amer. Math. Soc. **13** (1962), 706–711

[A–K] Alling, N. L.– Kuhlmann, S.: *On η_α-groups and fields*, Order **11** (1994), 85-92

[BO] Boshernitzan, M.: *Hardy fields and existence of transexponential functions*, Aequationes Mathematicae **30** (1986), 258–280

[BOU] Bourbaki, N.: *Commutative algebra*, Paris (1972)

[BR] Brown, R.: *Valued vector spaces of countable dimension*, Publ. Math. Debrecen **18** (1971), 149–151

[C] Cantor, G.: *Beiträge zur Begründung der transfiniten Mengenlehre*, Math. Ann. **46** (1895), 481–512

[CO] Collins, G. E.: *Quantifier Elimination for Real closed Fields by Cylindrical Algebraic Decomposition*, in: Automata Theory and Formal Language, 2nd G. I. Conference Kaiserslautern, Springer, Berlin (1975), 134–183

[C–K] Chang, C. C. – Keisler, H. J.: *Model Theory*, Amsterdam – London (1973)

[DA–G] Dahn, B. I. – Göring, P.: *Notes on exponential logarithmic terms*, Fund. Math. **127** (1986), 45–50

[DA–WO] Dahn, B. I. – Wolter, H.: *On the theory of exponential fields*, Zeitschr. f. math. Logik und Grundlagen d. Math. **29** (1983), 465–480

[DL–WD] Dales, H. G. – Woodin, W. H.: *Super–Real Fields; Totally Ordered Fields with Additional Structure*, London Math. Soc. Monographs, New Series **14**, Clarendon Press, Oxford (1996)

[D] van den Dries, L.: *A generalization of the Tarski-Seidenberg theorem, and some nondefinability results*, Bull. Amer. Math. Soc. **15** (1986), 189–193

[D1] van den Dries, L.: *Tame topology and o-minimal structures*, LMS Lecture Notes Series **248**, Cambridge University Press (1998)

[D2] van den Dries, L.: *T-convexity and tame extensions II*, J. Symb. Logic **62** (1997), 14–34

[D3] van den Dries, L.: *Remarks on Tarski's problem concerning $(\mathbb{R}, +, \cdot, \exp)$*, in: Logic Colloquium '82, eds. G. Lolli, G. Longo and A. Marcja, North-Holland (1984), 97–121

[D4] van den Dries, L.: *o-minimal structures*, in: Logic: From Foundations to Applications. European Logic Colloquium, eds. W. Hodges, M. Hyland, C. Steinhorn and J. Truss, Oxford (1996), 137-185

[D–L]   van den Dries, L. – Lewenberg, A. H.: *T-convexity and tame extensions*, J. Symb. Logic **60** (1995), 74–102

[D–M–M1]   van den Dries, L. – Macintyre, A. – Marker, D.: *The elementary theory of restricted analytic functions with exponentiation*, Annals Math. **140** (1994), 183–205

[D–M–M2]   van den Dries, L. – Macintyre, A. – Marker, D.: *Logarithmic-Exponential Power Series*, to appear in J. London Math. Soc.

[D–S1]   van den Dries, L. – Speissegger, P.: *The real field with convergent generalized power series is model complete and o-minimal*, to appear in Trans. Amer. Math. Soc.

[D–S2]   van den Dries, L. – Speissegger, P.: *The field of reals with multisummable series and the exponential function*, preprint

[F]   Fleischer, I.: *Maximality and ultracompleteness in normed modules*, Proc. Amer. Math. Soc. **9** (1958), 151–157

[FU]   Fuchs, L.: *Partially ordered algebraic systems*, Pergamon Press, Oxford (1963)

[G]   Gonshor, H.: *An introduction to the theory of surreal numbers*, Cambridge University Press (1986)

[GA]   Gabrielov, A.: *Projections of semi-analytic sets*, Functional Analysis and its Applications, **2** (1968), 282–291

[G–J]   Gillman, L.-Jerison, M. : *Rings of continous functions*, van Nostrand, Princeton (1960)

[GRA1]   Gravett, K. A. H.: *Valued linear spaces*, Quart. J. Math. Oxford (2), **6** (1955), 309–315

[GRA2]   Gravett, K. A. H.: *Ordered Abelian groups*, Quart. J. Math. Oxford (2), **7** (1956), 57–63

[H]   Hahn, H.: *Über die nichtarchimedischen Größensysteme*, S.-B. Akad. Wiss. Wien, math.-naturw. Kl. Abt. IIa, **116** (1907), 601–655

[HD1]   Hardy, G. H.: *Orders of Infinity*, Cambridge University Press (1910)

[HD2]   Hardy, G. H.: *Properties of Logarithmico-Exponential functions*, Proc. London Math. Soc. (2) **10** (1912), 54–90

[KA]   Kaplansky, I.: *Maximal fields with valuations I*, Duke Math. Journ. **9** (1942), 303–321

[KN–WR]   Knebusch, M. – Wright, M.: *Bewertungen mit reeller Henselisierung*, J. reine angew. Math. **286/287** (1976), 314–321

[KF1]   Kuhlmann, F.-V.: *Abelian groups with contractions I*, in: Abelian Group Theory and Related Topics, Proceedings of the Oberwolfach Conference on Abelian Groups 1993, (eds. R. Göbel, P. Hill and W. Liebert) Amer. Math. Soc. Contemporary Mathematics **171** (1994)

[KF2]   Kuhlmann, F.-V.: *Abelian groups with contractions II: weak o-minimality*, in: Abelian Groups and Modules. Proceedings of the Padova Conference 1994, eds. A. Facchini and C. Menini, Kluwer Academic Publishers, Dordrecht (1995)

[KF3]   Kuhlmann, F.-V.: *Valuation theory of fields, abelian groups and modules*, to appear in the "Algebra, Logic and Applications" series, eds. A. Macintyre and R. Göbel, Gordon & Breach, New York

[KS1]   Kuhlmann, S.: *On the structure of nonarchimedean exponential fields I*, Archive for Math. Logic **34** (1995), 145–182

[KS2]   Kuhlmann, S.: *Isomorphisms of Lexicographic Powers of the Reals*, Proc. Amer. Math. Soc. **123** (1995), 2657-2662

[KS3]   Kuhlmann, S.: *Valuation bases for extensions of valued vector spaces*, Forum Math. **8** (1996), 723–735

[KS4]   Kuhlmann, S.: *Infinitary Properties of valued and ordered vector spaces*, J. Symb. Logic **64** (1999), 216–226

[K–K1]   Kuhlmann, F.-V. – Kuhlmann, S.: *On the structure of nonarchimedean exponential fields II*, Comm. Alg. **22** (1994), 5079–5103

[K–K2]   Kuhlmann, F.-V. – Kuhlmann, S.: *The exponential rank*, The Fields Institute Preprint Series (1997), to appear in: Proc. of the Special Semester on Real Algebraic Geometry and Ordered Structures, Amer. Math. Soc. Contemporary Mathematics Series

[K–K3]   Kuhlmann, F.-V. – Kuhlmann, S.: *Residue fields of arbitrary convex valuations on restricted analytic fields with exponentiation I*, The Fields Institute Preprint Series (1996)

[K–K4]   Kuhlmann, F.-V. – Kuhlmann, S.: *Explicit construction of exponential-logarithmic power series*, preprint in: Structures Algébriques Ordonnées, Séminaire Paris VII (1997)

[K–K–S1] Kuhlmann, F.-V. – Kuhlmann, S. – Shelah, S.: *Exponentiation in Power Series Fields*, Proc. Amer. Math. Soc. **125** (1997), 3177–3183

[K–K–S2] Kuhlmann, F.-V. – Kuhlmann, S. – Shelah, S.: *Functional Equations in Lexicographic Products*, preprint

[K–P–S]   Knight, J. – Pillay, A. – Steinhorn, C.: *Definable sets in ordered structures II*, Trans. Amer. Math. Soc. **295** (1986), 593–605

[LAM1]   Lam, T. Y.: *The theory of ordered fields*, in: Ring Theory and Algebra III (ed. B. McDonald), Lecture Notes in Pure and Applied Math. **55** Dekker, New York (1980), 1–152

[LAM2]   Lam, T. Y.: *Orderings, valuations and quadratic forms*, Amer. Math. Soc. Regional Conference Series in Math. **52**, Providence (1983)

[LAN]   Lang, S.: *The theory of real places*, Ann. Math. **57** (1953), 378–391

[LAU1]   Laugwitz, D.: *Eine nichtarchimedische Erweiterung angeordneter Körper*, Math. Nachr. **37** (1968), 225–236

[LAU2]   Laugwitz, D.: *Tullio Levi-Civita's work on nonarchimedean structures*, in: Tullio Levi-Civita, Convegno internazionale celebrativo del centenario della nascita, Academia Nazionale dei Lincei, Atti dei Convegni Lincei **8** (1973)

[LC]   Levi-Civita, T.: *Sugli infiniti ed infinitesimi attuali quali elementi analitici (1892-1893)*, Opere mathematiche, vol. **1**, Bologna (1954), 1–39

[MI]   Miller, C.: *Exponentiation is hard to avoid*, Proc. Amer. Math. Soc. **122** (1994), 257–259

[M–MI]   Marker, D. – Miller, C.: *Levelled o-minimal structures*, Proc. of the Conference on Real Algebraic and Analytic Geometry, Revista Matemática, Servicio De Publicaciones Universidad Complutense (1997)

[MC–W]   Macintyre, A.J.-Wilkie, A.J.: *On the decidability of the real exponential field*, Kreisel 70th Birthday Volume, ed. P. G. Odifreddi, CLSI (1995)

[MP–M–S] Macpherson, D. – Marker, D. – Steinhorn, C.: *Weakly O-minimal structures*, preprint (1998)

[N] Neumann, B. H.: *On ordered division rings*, Trans. Amer. Math. Soc. **66** (1949), 202–252

[P–S] Pillay, A. – Steinhorn, C.: *Definable sets in ordered structures I*, Trans. Amer. Math. Soc. **295** (1986), 565–592

[PC] Prieß-Crampe, S.: *Angeordnete Strukturen. Gruppen, Körper, projektive Ebenen*, Ergebnisse der Mathematik und ihrer Grenzgebiete **98**, Springer (1983)

[PO] Poizat, B.: *Cours de théorie des modèles*, Nur el Mantiq wal Ma'arifa, Villeurbanne, (1985)

[PR1] Prestel, A.: *Lectures on Formally Real Fields*, Springer Lecture Notes in Math. **1093**, Springer, Berlin–Heidelberg–New York–Tokyo (1984)

[PR2] Prestel, A.: *Einführung in die mathematische Logik und Modelltheorie*, Vieweg studium, Braunschweig (1986)

[RE] Ressayre, J.-P.: *Integer parts of real closed exponential fields*, in: eds. P. Clote and J. Krajicek, Arithmetic, Proof Theory and Computational Complexity, Oxford University Press, New York (1993)

[RI] Ribenboim, P.: *Théorie des valuations*, Les Presses de l'Université de Montréal, Montréal, 1st ed. (1964), 2nd ed. (1968)

[RO] Robinson, A.: Function theory on some nonarchimedean fields, Amer. Math. Monthly **80** (1973), 87–109

[RO1] Rosenlicht, M.: *Hardy fields*, J. of mathematical analysis and applications **93** (1983), 297–311

[RO2] Rosenlicht, M.: *The rank of a Hardy field*, Trans. Amer. Math. Soc. **280** (1983), 659–671

[RO3] Rosenlicht, M.: *Growth properties of functions in Hardy field*, Trans. Amer. Math. Soc. **299** (1987), 261–271

[RO4] Rosenlicht, M.: *On Liouville's theory of elementary functions*, Pacific J. Math. **65** (1976), 485–492

[S] Shackell, J.: *Inverses of Hardy's L-functions*, Bull. London Math. Soc. **25** (1993), 150–156

[W1] Wilkie, A. J.: *Model completeness results for expansions of the ordered field of real numbers by restricted Pfaffian functions and the exponential function*, J. Amer. Math. Soc. **9** (1996), 1051–1094

[W2] Wilkie, A. J.: *A general theorem of the complement and some new o-minimal structures*, submitted (1997)

[Z1] Zil'ber, B.: *Generalized analytic sets*, Algebra i Logika, **36** (1997), 361–380 (in Russian)

[Z2] Zil'ber, B.: *Quasi-Riemann surfaces*, in: Logic: from Foundations to Applications, European Logic Coll., ed. W. Hodges, M. Hyland, C. Steinhorn and J. Truss, Clarendon Press, Oxford (1996), 515-536

Index

Additive Lexicographic Decomposition Theorem, 18
algebraic, 136
  χ-algebraic, 131
algebraic closure, 136
algebraic exchange property, 136
Alling, N., 27, 29
archimedean
  R–archimedean, 9
  equivalence class, 16
archimedean component
  corresponding to γ, 10
archimedean equivalent, 15
asymptotic, 96
automorphism
  increasing, 78
axioms
  n-th Taylor axiom, 33
  (C0), (C$\leq$), (C−), 120
  (CA), (CS), (CP), (CF), (CSN), (CZ), (CC), (CA′), 121
  (CE), 52
  (CO), 49
  (CV1), (CV2), 121
  (CV3), (CV4), (CV5), 122
  (EH1), (EH2), (EH3), 94
  (GA), 33
  (NV1), (NV2), (NV3), (NV4), 117
  (NV5), 118
  (OAG), 120
  (T_1), 53
  (T), 33
  (V0), (V1), (V2), (V3), (V4), 117
  growth axiom, 33
  Taylor axiom, 33

bounded
  χ-transcendental, 133
  exponentially, 95
  polynomially, 95
Brown's Theorem, 6
Brown, Ron, 6

canonical additive complement, 27
canonical group cross-section, 70
canonical multiplicative complement, 27
canonical valuation, 27

Cauchy sequence, 91
centrifugal, 46, 121
centripetal, 46, 121
chain, 1
characteristic domain, 138, 139
characteristic sequence, 131, 139
characterization
  of the exponential rank, 61
  of the principal exponential rank, 63
class
  R–equivalence class, 9
  archimedean, 16
  exponential, 58
  multiplicative, 52
closed
  χ_ℓ-closed, 61
  ℓ-closed, 61
  ζ_ℓ-closed, 61
  $\mathcal{F}$-closed, 97
  r$\mathcal{F}$-closed, 97
  exponentially closed, 32
  relatively exp-closed in, 108
coarser, 50
coefficient map corresponding to γ, 2
compatible, 23
  v-compatible exponential, 23
  v-compatible exponential field, 23
  v-compatible logarithm, 23
  w-compatible exponential, 52
  w-compatible prelogarithm, 53
compatible with the order, 9, 49
complement, 10
  canonical additive complement, 27
  canonical multiplicative complement, 27
component
  corresponding to γ, 2
  of x in M, 8
constant term, 27
contraction, 44, 121
  centrifugal, 46
  centripetal, 46
  natural, 45
  natural contraction induced by $\tilde{h}$, 45
  natural contraction induced by f, 45
contraction group, 121
contraction hull, 125

divisible, 125
convex, 8
  subgroup, 50
convex partition
  finite, 138
  finite open, 140
convex subgroup
  associated to w, 50
convex subring
  principal, 51
convex valuation, 16, 49
  associated to G_w, 50
Countable Case Characterization Theorem, 31
cross-section
  w-logarithmic, 66
  canonical group cross-section, 70
  group cross-section, 70
  of a valued field, 66
  strong v-logarithmic, 77
  surjective w-logarithmic, 66
Crucial Lemma, 104
cut, 118
  v_G-cut, 118
  induced by an element, 118
  shifted v_G-cut, 119

definable, 95
dense
  w-dense, 29
  order dense, 30
divisible, 16
  contraction hull, 125

elements
  finite, 17
  infinitesimal, 17
  positive infinite, 17
embedding
  truncation closed, 87
embedding of valued modules, 2
equivalent
  R–equivalent to y, 9
  φ-equivalent, 57
equivalent valuations , 2
exchange property
  algebraic, 136
exponential, 22
  T_1-exponential, 53
  v-compatible, 23
  v-left exponential, 24
  v-middle exponential, 25
  v-right exponential, 25
  w-compatible, 52
  (GA)-v-left exponential, 34
  (GA)-exponential, 33
  (GAT$_n$)-exponential, 33
  (GAT)-v-middle exponential, 34
  (GAT)-exponential, 33
  (T$_n$)-v-right exponential, 42

  (T$_n$)-exponential, 33
  (T)-v-right exponential, 34
  (T)-exponential, 33
  group exponential induced by χ, 45
  induced on the residue field, 52
  lexicographic product, 25
  strong group exponential, 35
exponential class, 58
exponential field, 22
  v-compatible, 23
  extension of exponential fields, 29
  formally, 22
exponential group in A, 26
exponential Hardy field, 94
exponential rank, 52
  characterization, 61
  of a prelogarithm, 53
  principal, 61
exponential-logarithmic power series field
  over Γ_0, 75
  over (Γ_0, σ), 81
exponentially bounded, 95
exponentially closed, 32
exponents
  field of, 97
extension
  immediate extension of valued modules, 3
  of exponential fields, 29
  of valued modules, 3
  value set preserving, 124

field of exponents, 97
final segment
  principal, 51
finer, 50
finite convex partition, 138
finite elements, 17
finite open convex partition, 140
formally exponential field, 22

germ at ∞, 92
Gravett, K. A. H., 12
group
  exponential group in A, 26
  of 1–units, 16, 18
  of positive units, 18
  of units, 16
  strong exponential, 35
group cross-section, 70
  canonical, 70
group exponential, 26
  induced by χ, 45
growth axiom, 33

Hölder's Theorem, 15
Hölder, O., 15
Hahn Embedding Theorem, 14
Hahn product, 3
Hahn sum, 3
Hardy field, 92

Index

exponential, 94
   levelled, 97
henselian, 28
homomorphism of ordered modules, 8

immediate extension
   of valued fields, 27
   of valued modules, 3
increasing automorphism, 78
induced by h_f, 26
infinitely smaller, 9
infinitesimals, 17
isomorphic as ordered skeletons, 10
isomorphic as valued modules, 2
isomorphism
   of ordered systems, 2
   of valued modules, 2

Kaplansky, I., 28

left
   v-left exponential, 24
   v-left logarithm, 24
   w-left prelogarithm, 66
   (GA)-v-left exponential, 34
level, 97
levelled, 97
lexicographic decomposition
   additive, 18
   multiplicative, 19
lexicographic product, 13
   of exponentials, 25
lexicographic sum, 13
lifting, 26, 66
lifting property, 35
lifts, 25
logarithm, 22
   v-compatible, 23
   v-left logarithm, 24
logarithmic cross-section, 66
   surjective, 66
Logarithmico-Exponential functions, 94

maximal exponential, 29
maximally valued
   field, 27
   module, 4
middle
   v-middle exponential, 25
   (GAT)-v-middle exponential, 34
minimum support valuation, 3, 27
monomial, 114
   generalized χ-monomial, 139
   monic generalized χ-monomial, 138
multiplicative class, 52
Multiplicative Lexicographic Decomposition
   Theorem, 19

natural contraction, 45
   induced by $\tilde{h}$, 45
   induced by f, 45

natural valuation
   R–natural valuation, 9
   on an ordered field, 16

o-minimal, 95
o-minimal, weakly, 136
open convex set, 139
order compatible, *see also* compatible with the order
order complete, 30
order completion, 30
order dense, 30
order preserving, 8
ordered module, 8
   homomorphism of, 8
ordered skeleton, 9
ordered system
   isomorphism of ordered systems, 2
   of modules, 2

piecewise monotone, 138
polynomial
   generalized χ-polynomial, 139
   reduced, 28
polynomially bounded, 95
positive infinite elements, 17
positive units, 50
power series field, 27
   exponential-logarithmic
      over Γ_0, 75
      over (Γ_0, σ), 81
precontraction, 56, 121
   centrifugal, 121
   centripetal, 121
precontraction group, 120
prelogarithm, 53
   w-compatible, 53
   w-left, 66
   w-right, 66
   (GA)-prelogarithm, 54
   (T_1)-prelogarithm, 55
   exponential rank, 53
principal
   χ_ℓ-principal, 61
   χ_ℓ-principal generated by g, 61
   ℓ-principal, 61
   ℓ-principal generated by a, 61
   ζ_ℓ-principal, 61
   ζ_ℓ-principal generated by γ, 61
   principal exponential rank, 61
principal convex subgroup
   generated by g, 51
principal convex submodule, 8
principal convex subring generated by a, 51
principal exponential rank, 61
   characterization, 63
principal final segment, 51
principal rank
   of a field, 51
   of a group, 51

pseudo Cauchy sequence, 5
pseudo complete, 5
pseudo limit, 5

quasicut, 118

rank
   σ-rank, 83
   exponential, 52
   of a valued group, 50
   of an ordered field, 50
   principal
      of a field, 51
      of a group, 51
   principal exponential rank, 61
realize a (quasi)cut, 118
reduced polynomial, 28
representable
   P-representable, 142
residue field, 16
restricted analytic function, 103
Riemann ζ-function, 113
right
   v-right exponential, 25
   w-right prelogarithm, 66
   (T_n)-v-right exponential, 42
   (T)-v-right exponential, 34
root closed for positive elements, 18

shift
   induced by a prelogarithm, 57
   to the right, 57
shift elements, 131
shift of a quasicut, 118
skeleton, 2
   isomorphic as ordered skeletons, 10
   ordered, 9
strong
   v-logarithmic cross-section, 77
   exponential group, 35
   group exponential, 35
sum of chains, 10
sum of ordered systems of modules, 10
supersequence, 131
support, 3, 27, 67
surjective w-logarithmic cross-section, 66

Taylor axiom, 33
theorem
   Additive Lexicographic Decomposition, 18
   Countable Case Characterization, 31
   Hölder's Theorem, 15
   Multiplicative Lexicographic Decomposition Theorem, 19
topology
   w-topology, 29
transcendental
   χ-transcendental, 131
   bounded χ-transcendental, 133
transexponential, 95
truncation closed embedding, 87

ultimately, 92
ultrametric inequality, 1
units
   group of 1–units, 16, 18
   group of positive units, 18
   group of units, 16
   positive, 50

valuation
   $v_{\min}$, 3, 27
   canonical, 27
   coarser, 50
   convex, 16, 49
   finer, 50
   on a field, 15
   on a module, 1
valuation basis, 4
valuation ideal, 16
valuation independent, 4
   over, 4
valuation preserving, 2
valuation ring, 16
value group, 15
value set, 1
value set preserving extension, 124
valued field, 15
   henselian, 28
   immediate extension, 27
   maximally valued, 27
valued module, 1
   embedding of valued modules, 2
   extension of valued modules, 3
   immediate extension of, 3
   isomorphism of valued modules, 2
   maximally valued, 4

weakly o-minimal, 136

List of Notation

$\bigoplus_{i \in I} M_i$ 1

v 1

(M, v) 1

$v(M)$ 1

$[\Gamma, \{B(\gamma) ; \gamma \in \Gamma\}]$ 2

$[\varphi, \{\varphi_\gamma \mid \gamma \in \Gamma_1\}]$ 2

M^γ 2

M_γ 2

$B(M, \gamma)$ 2

$S(M)$ 2

$B(\gamma)$ 2

$\pi^M(\gamma, x)$ 2

$\pi(\gamma, x)$ 2

$\prod_{\gamma \in \Gamma} B(\gamma)$ 3

$\operatorname{support}(s)$ 3

$\bigoplus_{\gamma \in \Gamma} B(\gamma)$ 3

$v_{\min}$ 3

$\coprod_{\gamma \in \Gamma} B(\gamma)$ 3

$\mathbf{H}_{\gamma \in \Gamma} B(\gamma)$ 3

${}^M_R \langle \{x_i \mid i \in I\} \rangle$ 4

$C_x(M)$ 8

$D_x(M)$ 8

$B_x(M)$ 8

$|x|$ 9

$\overset{R}{\sim}$ 9

$\overset{R}{\ll}$ 9

$[x]_R$ 9

Γ 9

v^R 9

List of Notation

$M \amalg N$ 10

$\Delta_1 + \Delta_2$ 10

$S_1 \amalg S_2$ 10

$<_l$ 13

(K, w) 15

v_G 16,117

R_w, I_w 16

Kw 16

bw 16

$\mathcal{U}_w$ 16

$1 + I_w$ 16

$K^{>0}$ 16

$[a]$ 16

$\overline{K}$ 17

$\mathbf{P}_K$ 17

v, v_G 17

$G^{<0}, G^{>0}, G^{\geq 0}$ 17

$\mathcal{U}_v^{>0}$ 18

$1 + I_v$ 18

$-v$ 19

$\dot{v}$ 19

$C_x, D_x, \mathbf{C}_x, \mathbf{D}_x, \mathcal{C}_x, \mathcal{D}_x$ 20

$\dot{\sim}, \dot{\gg}, \dot{\ll}$ 20

(K, f) 22

h_f 25

$f_L \amalg f_M \amalg f_R$ 25

$\tilde{h}_\ell$ 26

R^Γ 27

1_γ 27

$k((G))$ 27

$v_{\min}$ 27

$k[[G]]$ 27

$\text{Neg}(K)$ 27

$k((G^{<0}))$ 27

$\text{Mon}(K)$ 27

$p(x)w$ 28

Γ_K 30

List of Notation

K^c	30
$E_n(x)$	33
(GA)	33
(T)	33
(T_n)	33
$P_n(x,y), T'_n(x,y), Q_n(x,y)$	37
χ	44, 120
χ_f	45
R_w, I_w	49
(CO)	49
$\mathcal{U}_w^{>0}$	49
$\mathcal{R}$	50
G_w	50
Γ_w	51
Γ^{fs}	51
Γ^*	51
$\mathcal{R}^{\mathrm{pr}}$	51
$[a]$	52
(CE)	52
fw	52
$\mathcal{R}_f$	52
(T_1)	53
$\mathcal{R}_\ell$	53
χ_ℓ	56
ζ_ℓ	56
$\sim_\varphi$	57
$[a]_\varphi$	57
$\sim_\ell, \sim_f$	58
$\sim_{\chi_\ell}, \sim_{\zeta_\ell}$	58
$[a]_\ell$	58
$\mathbf{A}_w, \mathbf{B}_w$	65
$\ell_R^w, \ell w, \ell_L^w$	66
o-Emb$((1+I_w, \cdot), (I_w, +))$	66
h_ℓ^w	66
o-Emb$(w(K), (K, +) \setminus R_w)$	66
$\mathbf{L}_K, \mathbf{L}_K^w$	66
$\mathbf{X}_K, \mathbf{X}_K^w$	66

- support(a) 67
- $\mathbb{R}((\Gamma_0))^{EL}$ 75
- $\mathbb{R}((\Gamma_0))^{EL(\sigma)}$ 81
- $\log_n$, $\exp_n$ 94
- LE 94
- $\mathcal{R}$, $H(\mathcal{R})$ 95
- $f \sim g$ 96
- $\mathcal{P}$, $T(\mathcal{P})$ 97
- $\mathcal{F}_\pi$ 97
- κ 97
- $LE_\mathcal{F}(x)$ 98
- $\mathbb{R}_{\mathrm{an}}$ 103
- T_{an} 103
- $\mathcal{F}_{LE}$ 104
- $\mathcal{F}_{\mathrm{an}}$ 104
- $T_{\mathrm{an}}(\exp)$ 107
- u 107
- K_0^w, K_∞^w 109
- $LE_\mathcal{F}^w(\log_{m_0}(x))$ 110
- $\mathbb{R}((t))^{LE}$ 112
- $\mathcal{L}_{\mathrm{cg}}$ 120
- $\mathcal{P}_{\mathrm{cp}}$, $\mathcal{P}_{\mathrm{cf}}$ 123
- ρ_χ 124
- G_b 129
- Cut(b) 132
- $\mathcal{L}_{\mathrm{cg}}(G, x)$ 137
- $\mathcal{O}^a$, $\mathcal{M}^a$ 138
- $\chi[a]$, $\chi[a_1, \ldots, a_n]$ 139
- $G[a]$, $G[a_1, \ldots, a_n]$ 139
- $S_f(b)$ 151